Collins

Key Stage 3
Science

Student Book 1

Series editor: Ed Walsh
Authors: Sarah Askey
Tracey Baxter
Sunetra Berry
Pat Dower

William Collins' dream of knowledge for all began with the publication of his first book in 1819. A self-educated mill worker, he not only enriched millions of lives, but also founded a flourishing publishing house. Today, staying true to this spirit, Collins books are packed with inspiration, innovation and practical expertise. They place you at the centre of a world of possibility and give you exactly what you need to explore it.

Collins. Freedom to teach

Published by Collins
An imprint of HarperCollins*Publishers*
77 – 85 Fulham Palace Road
Hammersmith
London
W6 8JB

Browse the complete Collins catalogue at
www.collins.co.uk

10 9 8 7 6 5 4

ISBN 978-0-00-750581-4

British Library Cataloguing in Publication Data
A Catalogue record for this publication is available from the British Library

Commissioned by Letitia Luff
Project managed by Sally Moon, Jane Roth and Lyn Ward
Series editor Ed Walsh
Managing editor Caroline Green
Edited by Janette Edwards, Hugh Hillyard-Parker, Sally Moon, John Ormiston, Jane Roth, Lyn Ward and Ros Woodward
Proofread by Camilla Behrens and Tony Clappison
Editorial assistance by Lucy Roth
Designed by Joerg Hartmannsgruber
Cover design by Angela English
Picture research by Amanda Redstone
Illustrations by Jouve India Ltd and Ken Vail Graphic Design
Typesetting by Jouve India Ltd and Ken Vail Graphic Design
Production by Emma Roberts
Printed and bound by Grafica Veneta S.p.A., Italy.

Contents

How to use this book

These tell you what you will be learning about in the lesson.

This introduces the topic and puts the science into an interesting context.

Each topic is divided into three parts. You will probably find the section with the blue heading easiest, and the section with the purple heading the most challenging.

Try these questions to check your understanding of each section.

Chemistry

Finding the best solvent

We are learning how to:

- Choose the best solvent.
- Recognise hazards when using solvents.

Most graffiti artists use spray paints that cannot be washed away with water. To remove spray paint, a different solvent is needed to dissolve the paint.

FIGURE 1.3.15a: Some graffiti has become famous.

Solvent choice

There are many materials like spray paint and oil that are **insoluble** in water, so we need other **solvents** to **dissolve** them. For example, petrol is a very good solvent for oily stains and spray paint, but it is too dangerous to use because it is very flammable. Scientists use **hazard** symbols to highlight the risks of chemicals like solvents, see Table 1.3.15.

1. Why can we not simply wash away graffiti with water?
2. Which solvent would you choose to remove:
   a) nail polish from glass?
   b) ballpoint pen mark from a shirt?
   c) emulsion paint from paint brushes?
3. What hazards are involved when using solvents other than water?

FIGURE 1.3.15b: You can buy special stain removers.

TABLE 1.3.15: Properties of solvents

Solvent	It can dissolve	Hazards
water	sugar, food colours, emulsion paint	none
alcohol (ethanol)	ballpoint pen ink, perfume, herbs, spices	⬦ (flammable)
acetone (propanone)	nail polish	⬦ (flammable) ⬦ (harmful)
white spirit	grease, oil paint	⬦ (flammable) ⬦ (environmental) ⬦ (health hazard)

Careful choice

Tar stuck to the paint of a car is insoluble in water and difficult to remove. If you try to clean it off with solvents like acetone or white spirit, they will not only remove the tar, but also the paint on the car.

The choice of solvent is very important and must be selected by careful testing and checking. This is important for clothing too. If you use the wrong solvent you could damage the dyes and fabric.

4. What are the advantages and disadvantages of using tar and stone chippings on road surfaces?

5. Why might there be a problem in removing tar from a car?

6. Explain why there is a need to buy different stain removers when removing stains from clothing.

Clean and smelly

Dry cleaning uses a solvent called tetrachloroethene (C_2Cl_4) instead of water. The clothes are washed in the solvent at 30 °C before being tumbled in warm air (60 °C) to remove it. All the vapours produced are cooled and the condensed solvent is collected. Dry cleaners can recycle nearly 100 per cent of the solvent. This is important because the solvent is classified as highly toxic as well as harmful to the environment.

Alcohols like ethanol can dissolve colours, flavours and odours to make scented products like perfume. Alcohol evaporates easily, which is why perfume or aftershave dries so quickly on your skin. Liquids that evaporate quickly are described as **volatile**. This property allows us to smell substances, but it can also make solvents more dangerous. This is because the substances are more flammable and can enter the lungs quickly when they are vapours.

7. How is the dry-cleaning solvent recycled?

8. Why is it important that dry cleaners recycle as much solvent as possible?

9. What are the advantages and disadvantages of using a volatile solvent?

Did you know…?

Metals can dissolve in each other to form alloys. Dentists used to use mercury alloys for fillings in teeth. Now most fillings are made from ceramics that avoid the possible harmful effects of metal alloys in your mouth.

FIGURE 1.3.15c: Perfume evaporates easily.

Key vocabulary

insoluble

solvent

dissolve

hazard

volatile

Each topic has some fascinating extra facts.

These are the most important new science words in the topic. You can check their meanings in the Glossary at the end of the book.

Key this phrase into an internet search box to find out more.

How to use this book

The first page of a chapter has links to ideas you have met before, which you can now build on.

This page gives a summary of the exciting new ideas you will be learning about in the chapter.

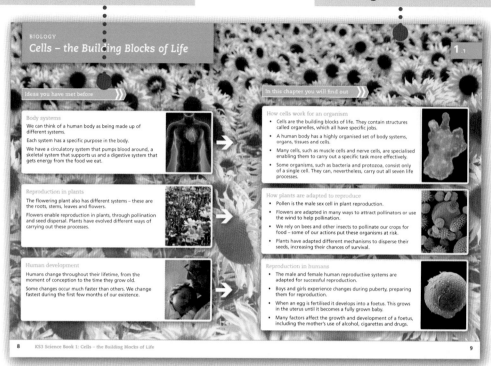

When you are about halfway through a chapter, these pages give you the chance to find out about a real-life application of the science you have been learning about.

The tasks – which get a bit more difficult as you go through – challenge you to apply your science skills and knowledge to the new context.

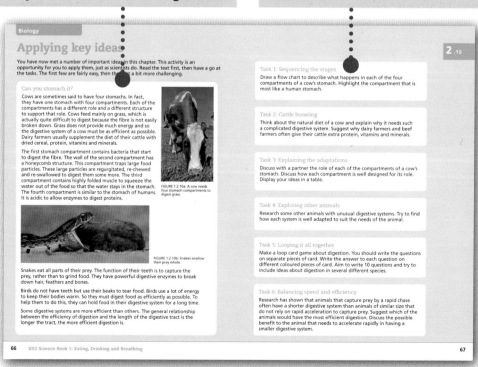

These lists at the end of a chapter act as a checklist of the key ideas of the chapter. In each row, the blue box gives the ideas or skills that you should master first. Then you can aim to master the ideas and skills in the orange box. Once you have achieved those you can move on to those in the purple box.

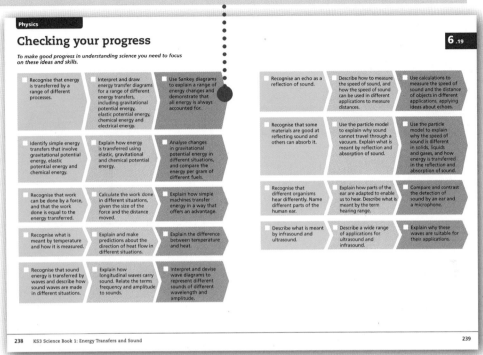

These end-of-chapter questions allow you and your teacher to check that you have understood the ideas in the chapter, can apply these to new situations, and can explain new science using the skills and knowledge you have gained.

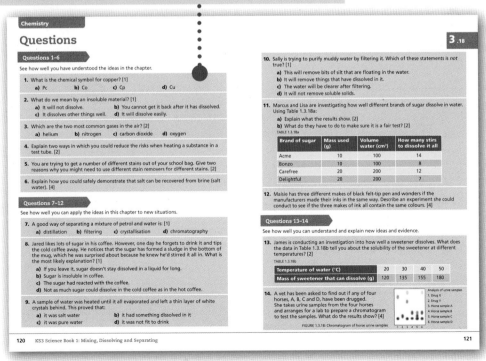

Cells – the Building Blocks of Life

Ideas you have met before »

Body systems

We can think of a human body as being made up of different systems.

Each system has a specific purpose in the body.

We have a circulatory system that pumps blood around, a skeletal system that supports us and a digestive system that gets energy from the food we eat.

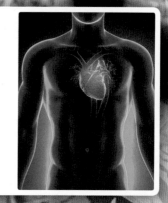

Reproduction in plants

The flowering plant also has different systems – these are the roots, stems, leaves and flowers.

Flowers enable reproduction in plants, through pollination and seed dispersal. Plants have evolved different ways of carrying out these processes.

Human development

Humans change throughout their lifetime, from the moment of conception to the time they grow old.

Some changes occur much faster than others. We change fastest during the first few months of our existence.

How cells work for an organism

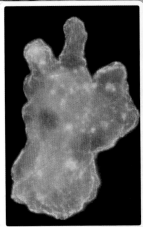

- Cells are the building blocks of life. They contain structures called organelles, which all have specific jobs.

- A human body has a highly organised set of body systems, organs, tissues and cells.

- Many cells, such as muscle cells and nerve cells, are specialised enabling them to carry out a specific task more effectively.

- Some organisms, such as bacteria and protozoa, consist only of a single cell. They can, nevertheless, carry out all seven life processes.

How plants are adapted to reproduce

- Pollen is the male sex cell in plant reproduction.

- Flowers are adapted in many ways to attract pollinators or use the wind to help pollination.

- We rely on bees and other insects to pollinate our crops for food – some of our actions put these organisms at risk.

- Plants have adapted different mechanisms to disperse their seeds, increasing their chances of survival.

Reproduction in humans

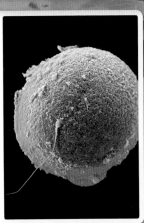

- The male and female human reproductive systems are adapted for successful reproduction.

- Boys and girls experience changes during puberty, preparing them for reproduction.

- When an egg is fertilised it develops into a foetus. This grows in the uterus until it becomes a fully grown baby.

- Many factors affect the growth and development of a foetus, including the mother's use of alcohol, cigarettes and drugs.

Historical ideas about living things

We are learning how to:

- Summarise some historical ideas about living things.
- Explain how evidence can change ideas.
- Select evidence to support or disprove ideas.

For many years people believed that living things came from non-living things. Today, water found in meteorites and moons suggests that life could have come to Earth from space.

FIGURE 1.1.2a: Meteorites like this may hold clues to the origins of life on Earth.

Spontaneous generation

From the time of Aristotle (384–322 BCE) to the 1600s, most people believed in the idea of spontaneous generation – that is, they thought that many **organisms** (living things) came from inanimate objects (non-living things). For example, observing mice coming out from a stack of corn, they would draw the **conclusion** that the corn had produced the mice.

1. Do the following observations seem to support or disprove the idea of spontaneous generation?

   a) kittens coming out of a barn

   b) fish swimming in a puddle

   c) lambs being born

2. Can you think of other examples where people might think that animals come from non-living things?

Redi's experiment

In 1668, Francesco Redi set out to disprove this idea. He put the same amount of fresh meat in three jars. He left one jar open, covered the other with a cheesecloth, and sealed the third. After a few days, maggots appeared in the open jar – there were no maggots in the closed jars. The maggots came from flies that had got into the open jar and laid eggs, not from the meat itself.

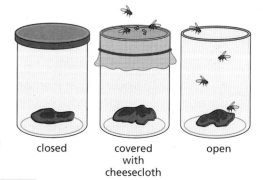

closed / covered with cheesecloth / open

FIGURE 1.1.2b: Redi's experiment to disprove the idea of spontaneous generation.

3. What scientific question was Redi trying to investigate?

4. How did Redi make his experiment a fair test?

5. How did his findings disprove the idea of spontaneous generation?

Disproving the idea ▶▶▶

In 1864, the scientist Louis Pasteur added the same amount of boiled broth to specially designed bottles. He sealed some bottles and removed the tops from the rest, then left them for a long time. He observed no life in any of the bottles that had been sealed, but the open bottles were teeming with life.

With the invention of the **microscope** in 1590, scientists observed that living things were complex structures, which could not have possibly been formed from inanimate objects. From studying samples of cork bark, Robert Hooke discovered that organisms were made from simple building blocks. We call these individual building blocks **cells**. They are too small to be seen with the unaided eye.

> **Did you know…?**
>
> Today, scientists can use special high-power microscopes to study structures within cells to find cures for diseases. Modern-day microscopes are also used to study the structure of crystals and metals. Some can even view atoms.

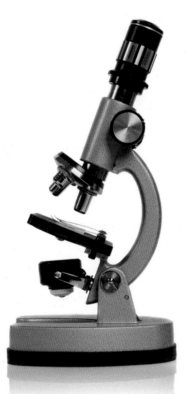

1938 Ernst Ruska develops the electron microscope to improve the magnification and resolution. Viruses and molecules are studied

▲

1932 Frits Zernike invents a microscope to study transparent and colourless specimens

▲

18th Century improvements in microscopes result in their greater use by scientists

▲

1675 Anton van Leeuwenhoek uses a simple microscope to look at blood, insects and pond water. He was the first person to describe cells and bacteria

▲

1667 Robert Hooke makes a microscope to study various objects

▲

1590 Dutch lens grinders Hans and Zacharias Jansen make the first microscope by placing two lenses in a tube

FIGURE 1.1.2c: The invention of the microscope enabled the discovery of cells.

6. What conclusions would you reach based on the **evidence** from Pasteur's experiment?

7. Which investigation would you trust the most – Pasteur's or Redi's? Give a reason for your answer.

8. Why did it take so long for people to change their ideas after Redi's investigation?

9. What impact do you think microscopes have had on our understanding of living things?

Key vocabulary

organism

conclusion

microscope

cell

evidence

Comparing plant and animal cells

We are learning how to:

- Develop models to explain the differences between animal cells and plant cells.
- Record evidence using a microscope.
- Communicate ideas about cells effectively using scientific terminology.

Every cell is a chemical processing factory, with over 500 quadrillion chemical reactions occurring every second! Without these reactions, the organism would die.

Cells as building blocks

All living organisms are made of cells – they are the building blocks of life. Cells cannot be seen except under a microscope. This is why it took so long to discover them. Some organisms are made of only one cell; most are made of millions of cells working together.

1. How can we see cells?

2. Is a cell living?

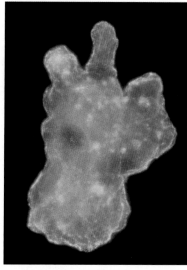

FIGURE 1.1.3a: An amoeba is a single-celled organism.

Common structures in animal and plant cells

All plant cells and animal cells have three main structures – the **nucleus**, the **cytoplasm** and the **cell membrane**.

Every cell, except red blood cells, contains a nucleus. The nucleus contains DNA, which controls the reactions inside the cell and is involved in making the cell reproduce.

The cytoplasm is a jelly-like material that makes up the bulk of the cell. All the chemical reactions occur here. Smaller structures within the cytoplasm, called organelles, make new materials to keep the cell and the organism alive.

The cell membrane surrounds the cell and contains the cytoplasm. The cell needs water, oxygen, glucose and nutrients – the membrane lets these in. During the chemical reactions, the cell makes waste products that it must get rid of, including carbon dioxide and urea. The membrane lets these substances out of the cell.

In the cytoplasm, special organelles called **mitochondria** convert glucose and oxygen into a form of energy that the cell can use.

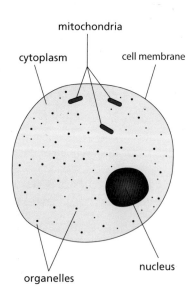

FIGURE 1.1.3b: The main structures of an animal cell

3. Which two parts of the cell are found inside the cytoplasm?

4. What main substances are allowed through the cell membrane?

Differences between animal and plant cells 〉〉〉〉

Animal cells are the simplest type of cell, containing a nucleus, cytoplasm, a cell membrane and mitochondria in the cytoplasm. Plant cells share these parts, but also have other important structures.

The **cell wall** is an extra protective layer outside the cell membrane. It gives the cell shape and strength.

The **vacuole** is a large bubble full of liquid, storing water, sugars, nutrients and salts in the cytoplasm. It provides internal pressure for the cell, keeping it firm and in shape. It also helps to control water movement inside and between cells.

Leaf cells also contain small, round, green organelles called **chloroplasts**. These contain a green pigment called chlorophyll, which absorbs energy from the Sun and helps the plant make glucose.

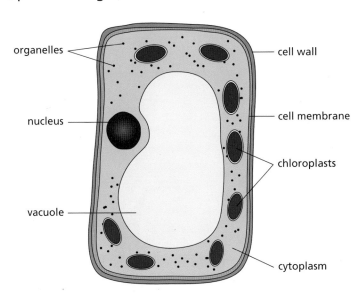

FIGURE 1.1.3c: A plant leaf cell

Did you know…?

Each human consists of about 100 trillion cells working together.

Key vocabulary

nucleus

cytoplasm

cell membrane

mitochondria

cell wall

vacuole

chloroplast

5. Which two structures give a plant cell its shape?

6. Which cell do you think will be larger – a plant cell or an animal cell? Explain your answer.

7. Why do you think plant cells need extra structures that are not found in animal cells?

Describing cells

We are learning how to:

- Classify specialised cells as animal cells or plant cells.
- Describe different specialised animal cells and plant cells.
- Explain the structure and function of specialised cells using models.

All young cells start out exactly the same – these are called stem cells. When they grow, stem cells change their structure to carry out a certain job within the organism. Any stem cell can be made to become any type of specialised cell.

The right cells for the job

Many animal cells look very different from each other, although they contain the same basic structures. Cells become *specialised* so they can carry out a particular job. In an organism, many different jobs need to be done to keep it alive. These include movement, detecting information about the environment, sending messages, carrying chemicals around the body, making chemicals the body needs, reproducing and absorbing food.

1. Where would you find cells that detect:

    a) light?   b) sound?   c) heat?

Specialised animal cells

Nerve cells have very long extensions of cytoplasm. This enables them to carry messages from one part of the body to another.

Muscle cells are made from protein fibres that can rapidly expand and contract to create movement. They have the most mitochondria of all cells because they need lots of energy.

Sperm cells have tails and huge heads. Their main job is to carry genetic material to an **egg cell**, so that it can be fertilised. Sperm cells have lots of mitochondria because they must swim long distances.

2. Name the animal cells in Figure 1.1.4a.

3. Which cell:

    a) transmits electrical messages?

    b) contracts and expands to create movement?

    c) carries genetic material for fertilisation?

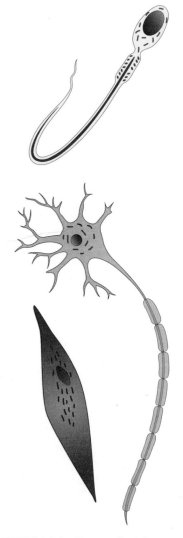

FIGURE 1.1.4a: Can you find the nucleus, cell membrane, cytoplasm and mitochondria of each cell?

Plant cells are also highly specialised. Plants make their own food by a process called photosynthesis. Many of the specialised cells in a plant are linked to this function. Cells collect light and water, and take in carbon dioxide. Specialised leaf cells like the one shown in Figure 1.1.3c in Topic 1.3, use these materials and turn them into sugar.

Specialised plant cells are also linked to the process of reproduction. Pollen cells are the male sex cell in plants. Some are carried by the wind, and others stick to insect or bird pollinators. There are over 300 000 different types of pollen cells.

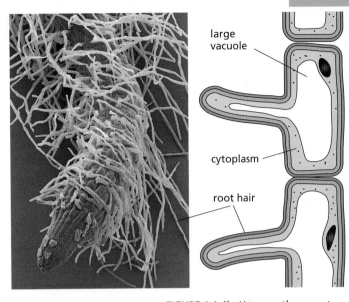

large vacuole

cytoplasm

root hair

FIGURE 1.1.4b: How are these root hair cells different from the leaf cell shown in Figure 1.1.3c?

4. Look at Figure 1.1.3c showing a leaf cell and at Figure 1.1.4b showing a **root hair cell**. Describe the features of each and suggest how these features enable the cells to carry out their jobs.

5. Compare and contrast the specialisation of a wind-transferred pollen cell and an insect-transferred pollen cell. Look at Figure 1.1.4c to help you.

6. Which is more specialised – a pollen cell or a sperm cell? Give reasons for your answer.

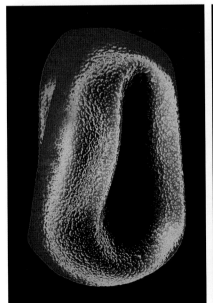
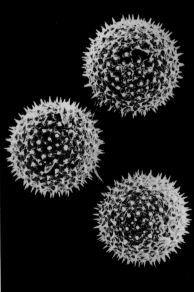

FIGURE 1.1.4c: Wind- and insect-transferred pollen cells.

Did you know…?

There are more than 200 different types of specialised cells in the body. In 2012, a Nobel Prize was awarded for the discovery that specialised cells can be changed to become stem cells.

Key vocabulary

nerve cell

muscle cell

sperm cell

egg cell

root hair cell

Understanding unicellular organisms

We are learning how to:

- Recognise different types of unicellular organisms.
- Describe differences in unicellular organisms.
- Compare and contrast features of unicellular organisms.

The oldest unicellular organisms were found in rocks dated to 3.8 billion years ago. They used chemicals in the ocean for 'food'. Around 3.5 billion years ago, organisms that could make their own food also evolved. Unicellular organisms were the main form of life on the planet for nearly 2 billion years.

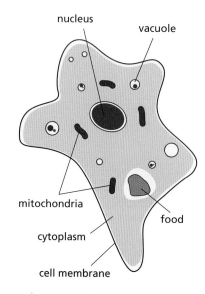

FIGURE 1.1.5a: An amoeba can carry out all life processes.

Unicellular organisms 〉〉

Unicellular organisms are made up of just one cell. They carry out all the life processes needed to exist independently. They differ from each other in their structure, how they feed and how they move. Algae are plant-like unicellular organisms containing chloroplasts and make their own food. Animal-like unicellular organisms take in food through their cell membrane. Some have developed tiny hairs to help them move, so they can find food or escape from predators. Some are themselves predators and will devour other unicellular organisms. Fungus-like unicellular organisms are called **yeasts**. They have a cell wall but cannot make their own food.

1. Name three different unicellular organisms.

2. List three ways unicellular organisms differ from each other.

Prokaryotes 〉〉〉

Unicellular organisms can be classified into two main groups – **prokaryotes** and **eukaryotes**. Prokaryote means 'before life' – prokaryotes are thought to be the first organisms to live on Earth. They do not have a nucleus, and their genetic material floats within the cytoplasm. They can be up to 200 times smaller than eukaryotes. **Bacteria** are examples of prokaryotes. They come in different shapes and sizes, live in different environments and have a range of food sources. Some bacteria take in chemicals from their environment, such as iron and sulfur, and use these as food. Others contain chloroplasts and use sunlight to make their own food – many

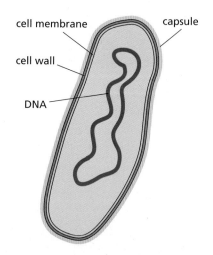

FIGURE 1.1.5b: A bacterium has no nucleus and no mitochondria.

can absorb nutrients from their environment. Bacteria can be found in extreme conditions, from under-sea volcano vents to places with temperatures well below freezing.

3. Look at Figure 1.1.5a and Figure 1.1.5b. Which is a prokaryote and which is a eukaryote?

4. What differences can you see between prokaryotes and eukaryotes?

Eukaryotes 》》》》

Eukaryotes contain a nucleus, surrounded by a nuclear membrane. They also contain many organelles (which prokaryotes do not), including mitochondria, chloroplasts and vacuoles. Examples of eukaryotes are euglena (a type of algae containing chloroplasts), yeast, amoeba, and paramecium – the last two are types of **protozoa**. Eukaryotes can be up to 200 times bigger than prokaryotes and often have external features to help them to survive. The amoeba can move around because its cytoplasm can flow; paramecium has cilia that beat and enable it to move, and the euglena has a flagellum, or tail, to enable it to move.

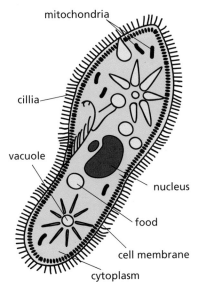

FIGURE 1.1.5c: A paramecium

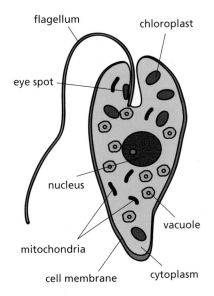

FIGURE 1.1.5d: Euglena

Did you know…?

Nummulites are the largest known unicellular organisms. Nummulite fossils as large as 16 cm in diameter have been found, which is about the size of a tennis ball. Some are thought to have lived for over 100 years.

Key vocabulary

yeast

prokaryote

eukaryote

bacterium

protozoa

5. Look at Figure 1.1.5d. How does euglena get its food?

6. Which is the most effective form of movement between the three eukaryotes? Justify your choice.

7. Summarise, in a table, the main similarities and differences between unicellular organisms.

Understanding diffusion

We are learning how to:

- Describe the process of diffusion and its relation to the cell.
- Plan a fair test investigation to explore the factors affecting diffusion.
- Explain how the different factors speed up or slow down diffusion.

How do substances move from the outside of a cell to the inside of a cell? One answer lies in the process of diffusion. Using this, and other processes, cells allow only the substances they need to enter the cell, and keep themselves safe from unwanted and toxic chemicals.

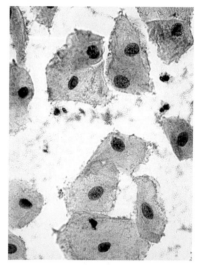

FIGURE 1.1.6a: Substances need to pass in and out of the cell membrane.

Chemicals on the move »

All cells require chemicals including **glucose**, oxygen, nutrients and minerals in order to survive. These pass through the cell membrane by **diffusion**, a process by which substances move from an area of high concentration to one of low concentration, until the concentrations are equal. There have to be more particles outside the cell than inside for them to move into the cell.

Cells also produce waste products, such as carbon dioxide and urea. They move out of the cell by diffusion because there are more waste particles inside the cell compared to outside.

1. Look at Figure 1.1.6b. Which way will the particles move?

2. Draw an outline of a cell showing the movement of named substances into and out of the cell.

Factors affecting diffusion »»

Many factors affect how quickly diffusion occurs. It occurs more rapidly at higher temperatures because the particles have more energy and move faster.

The number of particles in a given volume is called the concentration of a solution. In a highly concentrated solution, there are many particles packed into a small space. The particles try to move away from each other as quickly as possible. If a cell is placed in a high concentration of nutrients, the nutrients diffuse faster into the cell compared to when it is in an area of low concentration.

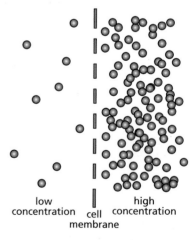

low concentration   high concentration

cell membrane

FIGURE 1.1.6b: Particles have a different concentration on either side of the cell membrane.

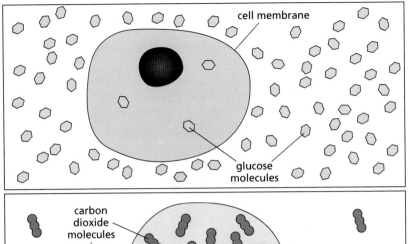

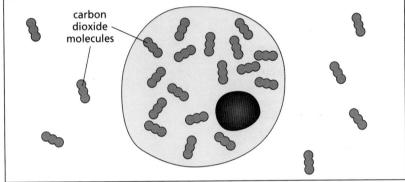

FIGURE 1.1.6c: Diffusion will occur in these 'cells'. Why will this happen?

3. Draw two identical boxes. Using small circles to represent particles, show a concentrated solution in one and a dilute solution in the other.

4. Suggest and explain whether temperature or concentration has a greater effect on the rate of diffusion.

Effect of surface area >>>

The survival of unicellular organisms depends on the rate of diffusion of chemicals into and out of the cell. The ratio of the **surface area** of the cell to its **volume** affects how quickly diffusion across the cell membrane can occur. With a higher surface-area-to-volume ratio, the rate of diffusion into and out of the cell is faster.

The surface area of a cube can be calculated by working out the area of one side and multiplying this by the number of faces. The volume of a cube is its length × breadth × height. The surface-area-to-volume ratio is worked out by dividing its surface area by its volume.

5. What can you say about the ratio of the surface area to volume as the size of cells increases?

6. Why are most unicellular organisms microscopic?

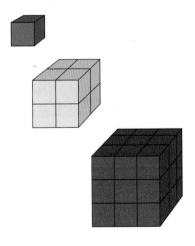

FIGURE 1.1.6d: These cubes represent cells of different sizes.

Key vocabulary

glucose

diffusion

surface area

volume

Understanding organisation in multicellular organisms

We are learning how to:

- Define the terms tissues, organs and organ systems.
- Explain the organisational structure in multicellular organisms.
- Compare the strengths and weaknesses of multicellular organisms and single-celled organisms.

The first simple multicellular organisms are thought to have evolved about 1.2 billion years ago. These eventually increased in organisation and size to form complex multisystem, multicellular organisms. There are 15 different organ systems within human beings, all working together to help us to survive.

Cells, tissues and organs

Groups of similar specialised cells working together are called **tissues**. Examples of human tissues are muscles and bones. Different tissues work together to make up an **organ**. Every organ has a specific job – the eye is an organ made up of many different tissues including a lens and an iris. They work together to enable us to see. Examples of other organs are:

- the heart, which pumps blood to the cells
- the kidneys, which clean the body and balance water in the body
- the brain, which allows us to control all parts of our body quickly.

Organs work together to make **organ systems**. Some of the organ systems in the human body are the circulatory system, the skeletal system, the respiratory system, the digestive system, the reproductive system and the nervous system.

muscle tissue

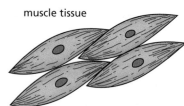

leaf tissue

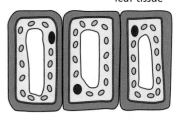

FIGURE 1.1.7a: Two types of tissue as seen under a microscope. What do you notice about the cells in each type of tissue?

1. Name three other organs and describe their functions.

2. The skin is described as an organ, not a tissue. Suggest why.

3. a) Name an organ in each of the six organ systems listed in the text.

   b) State the function of each of these organ systems.

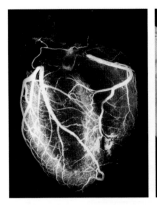
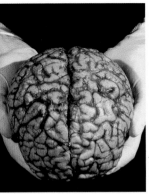
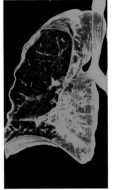

FIGURE 1.1.7b: What do all of these have in common?

Unicellular organisms

Unicellular organisms can live independently of other cells. They have no organ systems, just organelles working together. Because of their small size, they can reproduce very quickly. However, because they are unicellular, there is a limit to how big they can grow and most remain microscopic. If they become too big they cannot obtain the chemicals they need from the environment quickly enough by diffusion. However, being small means they are vulnerable to attack from bigger unicellular, or multilcellular, organisms.

4. Unicellular organisms make up most of the mass of biological material on the planet. What does this say about the success of unicellular organisms?

5. Is being small an advantage or a disadvantage? Give reasons.

How cell types evolved

Some cells evolved to join and work together, forming colonies of cells. An advantage of this was that in times of food shortage, food could be caught, digested and shared more effectively by cells working together. Eventually some of the cells within the colonies became specialised and took on particular jobs. This eventually led to the formation of simple multicellular organisms. These could grow to be much larger than the unicellular organisms and so were better protected and could move further in search of food. However, they needed to evolve complex organ systems in order to become much larger. This requires a lot of energy, and the larger the organism became, the slower the rate of reproduction.

6. What are the advantages of multicellular organisms over unicellular organisms?

7. Why did some cells form colonies?

Did you know…?

Many unicellular organisms live as parasites within multicellular organisms. Most are harmless, but some cause diseases such as malaria and typhoid. The number of bacterial cells living in our gut and on our skin is bigger than the number of our own cells.

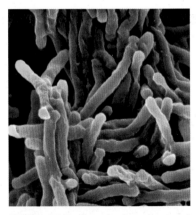

FIGURE 1.1.7c: How do these bacterial cells survive better by working together?

Key vocabulary

tissue

organ

organ system

Applying key ideas

You have now met a number of important ideas in this chapter. This activity gives an opportunity for you to apply them, just as scientists do. Read the text first, then have a go at the tasks. The first few are fairly easy – then they get a bit more challenging.

The skin is an organ

The skin is the largest organ in the human body. It is our first line of defence against heat, light, injury, and infecting bacteria and fungi. It also protects us from harmful radiation from the Sun, which can cause cancer. Skin cancer is the most common type of cancer, with over a million new cases reported every year worldwide.

The skin is about 2 mm thick and is composed of three different layers of tissue.

The top layer of skin tissue is called the epidermis. These cells are lost regularly and are replaced every six to eight weeks. We lose about 30 000 to 40 000 skin cells every hour!

The middle layer of tissues is called the dermis. This contains blood vessels, nerve cells and elastic tissue called collagen, which keeps the skin from sagging.

The bottom and thickest layer of tissue is the hypodermis. This layer is responsible for storing fat cells.

Specialised cells in the skin perform different jobs:

- cells that collect information about heat, pain and pressure
- cells that store fat to keep us warm
- pigment-containing cells that protect us from harmful rays from the Sun
- hair follicles that are useful in controlling temperature
- sweat glands that also help to control the body's temperature.

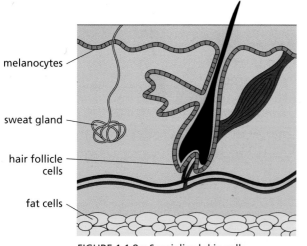

FIGURE 1.1.8a: Specialised skin cells

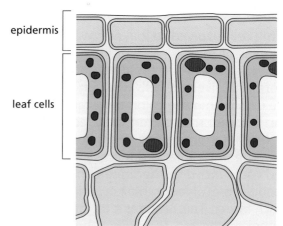

FIGURE 1.1.8b: Plant epidermis

In plants the epidermis is just one cell thick. It is a waterproof layer with a waxy coating. There are far fewer specialised cells here compared with animal skin.

Task 1: Animal cells

Draw a labelled diagram of a non-specialised animal cell in the epidermis of the skin.

Task 2: Plant cells

Draw a labelled diagram of a non-specialised plant cell in the epidermis of a plant. Describe how the animal cells and plant cells are alike and and how they are different.

Task 3: Organisation in multicellular organisms

The skin is an organ composed of tissues and specialised cells. Giving examples, explain what is meant by the terms 'organ', 'tissue' and 'specialised cell'.

Task 4: Specialised cells

Explain the features of two specialised cells in human skin. Explain how these features allow them to carry out their jobs.

Task 5: Unicellular organisms

Impetigo is a skin condition caused by bacteria. Athlete's foot is another skin disorder, caused by a fungus. Use diagrams to explain the differences between bacteria and fungi.

Why might it be hard for paramecium or euglena to live in the epidermis?

Comparing flowering plants

We are learning how to:

- Describe the structures and functions of parts in flowering plants.
- Explain why different plants have such diverse structures.
- Evaluate the differences between wind-pollinated plants and insect-pollinated plants.

The first plants on Earth were mosses. These relied on moisture and touch to transfer pollen. The first flowering plants, using wind and insects to transfer pollen, are thought to have evolved about two-hundred million years ago. Nowadays about 70 per cent of plant species use insects, birds or mammals to transport pollen.

Flowers as reproductive organs

Most flowers have male and female parts. The male part is the stamen, consisting of an **anther** and a **filament**. The anther produces **pollen**, which contains the male sex cell. The female part is the carpel. This consists of an **ovary** (with the female sex cells in the ovules), the **style** and the **stigma**, which has a sticky top. The purpose of the flower is to produce pollen in the anther and transfer it to the stigma of a different plant. This process, called pollination, is mainly achieved using wind, insects, birds or bats.

FIGURE 1.1.9a: Which flower is wind pollinated and which is insect pollinated?

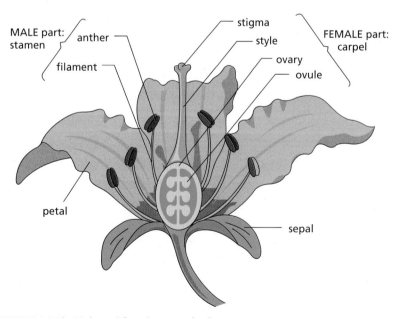

MALE part: stamen — anther — filament

stigma — style — ovary — ovule

FEMALE part: carpel

petal

sepal

FIGURE 1.1.9b: Male and female parts of a flower

1. Identify the following parts in the photos in Figure 1.1.9a: anther, filament, stamen, stigma, style, ovary.

2. What differences can you see between the two flowers in Figure 1.1.9a?

3. Why do you think the flowers have the differences you have written in your answer to question 2?

Attracting insects

Most insect-pollinated plants produce brightly coloured flowers with sweet smells to attract insects. Many also produce nectar deep inside the flower. This is a sugary fluid that draws insects inside the flower to encourage pollination. Pollinators such as bees collect the pollen as a food source. Plants produce a lot of pollen to increase the chances of successful pollination. Some orchids produce flowers that look like the female of particular wasps.

FIGURE 1.1.9c: This orchid mimics the appearance of a female wasp. The male wasp visits the flower and becomes covered in pollen.

4. Describe different ways plants encourage insects to visit them.

5. Why do plants use such a diverse range of methods of attracting pollinators?

Wind or insect pollination?

There is no guarantee that the wind will successfully transfer the pollen from one plant to the stigma of another plant, so wind-pollinated plants produce millions of pollen cells to improve their chance of success, even though most cells are wasted. Stigmas evolved to become large and feathery so as to capture pollen floating on the wind. Even so, there is no guarantee that the pollen from the same species will land on the plants.

Insect-pollinated plants produce far less pollen, but use other mechanisms to attract insects. However, some insects eat parts of the flower and plant, so flowers have developed mechanisms to avoid this, such as producing toxins and growing spikes.

6. Discuss the advantages and disadvantages of wind pollination and insect pollination.

Did you know…?

The oldest-known pollen grains were found on the bodies of tiny insects encased in amber. The pollen was thought to be over a million years old. Fossilised pollen has provided evidence of how plant life on Earth has evolved.

Key vocabulary

anther

filament

pollen

ovary

style

stigma

Knowing how pollination leads to fertilisation

We are learning how to:

- Describe the processes of pollination and fertilisation.
- Analyse and present data on the growth of pollen tubes.
- Explain factors that affect the growth of pollen tubes.

The world's chocolate supply depends on midges. These tiny flies are the only insects that can pollinate the cacao plant. Once fertilised, the plant produces seeds, which are used to make coffee and chocolate.

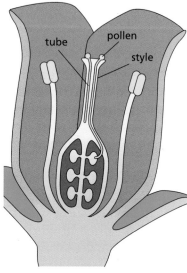

FIGURE 1.1.10a: How do you think a pollen tube is formed, from one cell or many?

Fertilising plants

Pollen contains the male sex cell. The female sex cell, the egg cell, is found in the ovule. The pollen travels down the style to reach the ovule by growing a long tube. Once it has reached the ovule, the nucleus of the male sex cell joins with the nucleus of the egg cell – this is **fertilisation**. The result is a new seed, which will eventually become a new plant.

1. Describe how pollination and fertilisation differ.

Pollen tubes

Look at Figure 1.1.10b. When a pollen grain lands on the stigma of another plant, the tube cell uses stored nutrients and sugars to grow a **pollen tube** down to the ovule. The concentration of sugar affects the ability of the pollen grains to grow tubes.

2. Plot a graph of the data in Table 1.1.10a.

3. Describe the pattern shown by the data.

TABLE 1.1.10a: The effect of sugar on the growth of pollen tubes

Sugar concentration (%)	5	10	15	20
growth of pollen tubes (micrometres)	250	350	450	200

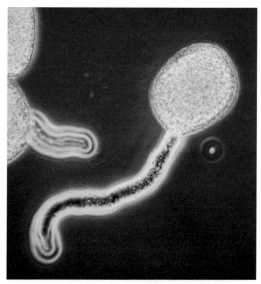

FIGURE 1.1.10b: Why might pollen from different species grow best in different concentrations of sugar solution?

Many factors affect the growth of pollen tubes. The pollen grain is dry when it lands on the stigma. Chemicals in the stigma enable water to enter the pollen grain so it can grow. These include sugar. In some plants, if pollen lands on the stigma of the same plant, chemicals prevent it from growing pollen tubes so it cannot fertilise itself. If the temperature is hotter, pollen tubes grow faster. Once the tube has formed, chemicals in the stigma direct the pollen tube to the ovule. Only one pollen cell fertilises the egg.

4. Look at the data in Table 1.1.10b and highlight any anomalous results.

TABLE 1.1.10b: The growth of pollen tubes in different sugar concentrations

Sugar concentration (%)	5	10	15	20
Growth of pollen tubes (micrometres) – experiment 1	225	345	200	213
Growth of pollen tubes (micrometres) – experiment 2	250	350	450	207
Growth of pollen tubes (micrometres) – experiment 3	275	355	450	250

5. Ignoring the anomalous values, calculate the average values in Table 1.1.10b.

TABLE 1.1.10c: The effect of temperature on pollen tube growth

Temperature (°C)	15	20	25	30	35	40
Growth of pollen tubes (micrometres)	0	200	420	700	800	100

6. From the data in Tables 1.1.10b and 1.1.10c, draw two graphs – one showing the effect of sugar concentration and the other showing the effect of temperature on pollen tube growth.

7. Which has the bigger effect on pollen tube growth, from these data? Why do you think this?

8. Which data would you trust more – explain why.

Did you know…?

Pollen that does not land on a stigma remains in the environment. It is the primary cause of hay fever and allergies. Pollen counts are made by counting how much pollen lands on a greasy spinning rod over a 24-hour period.

Key vocabulary

fertilisation

pollen tube

Understanding the challenges facing pollinators

We are learning how to:

- Describe the role of insects in crop production, using suitable data.
- Explain why bee populations are declining.
- Make suggestions for increasing insect populations, and hence crop production.

Honey bees used to pollinate 70 per cent of the UK's insect-pollinated crops. Today, it is less than 30 per cent. However, the growth of some crops has risen, suggesting that other insects have been taking the place of honey bees. Pollinators must be protected to make sure we don't lose crops.

Important pollinators

Most cereal crops are wind-pollinated. However, insect, bird and bat pollinators are responsible for 35 per cent of global crop pollination, including pollination of fruits, nuts, seeds, beans, coffee, oilseed rape, onions, almonds and tomatoes. Butterflies and bees are among the most important pollinators. Bees collect pollen from plants to make honey. They are usually specially **adapted**, with a furry body, so they can collect the maximum amount of pollen. Some of this lands on other plants as they move from plant to plant.

1. Name three other crops we rely on insects to pollinate.

2. Why are bees so useful as pollinators?

3. If a plant relies on one species of pollinator, what will happen if the pollinator dies out?

FIGURE 1.1.11a: The world's most important pollinator

Confused bees

Since 2005, more than ten million bee colonies have been wiped out by **colony collapse disorder (CCD)**, possibly caused by **pesticides**. Bees become weak and confused, and can't find their way back to the hive. This results in a reduction in the size of the colony, a shortage of food for the remaining bees and the inability to reproduce successfully. Many bees have also shown signs of increased viral disease.

4. Draw a flow chart to show how CCD occurs and its effect on a bee colony.

Theories about the causes of CCD include:

- Pesticides such as **insecticides** are sprayed on crops to prevent insects and fungi from attacking them. Unfortunately it is not possible to target specific insects, and others may be affected. Some contain nicotine, which is thought to cause confusion in the brain of the bee.

- The number of wild flowering plants has reduced as more land is used for development and agriculture, leading to less variety of pollen and a narrower range of nutrients for bees.

- Farmers rent out hives to pollinate crops. This can disorientate bees, which find their way around by locating routes back to their hives. Also, disease can be spread more widely because hives from different locations come into close contact, which would not occur naturally.

- Climate change means that some plants are flowering earlier, before bees can fly.

- Bees are more susceptible to virus attacks because other factors have made them weaker.

> **Did you know…?**
>
> In the USA, bees are the only pollinators of almond trees. Farmers are completely dependent on an active bee population for a good crop.
>
>
>
> FIGURE 1.1.11c

FIGURE 1.1.11b: What are the problems with industrial pollination?

5. What do you think is the most likely cause of CCD?

6. Describe three steps that could be taken that would increase bee **populations**.

Key vocabulary

adapted

colony collapse disorder (CCD)

pesticide

insecticide

population

Understanding how seeds are dispersed by the wind

We are learning how to:

- Recognise the variety of different structures shown by different seeds.
- Describe the need for plants to disperse their seed.
- Plan an investigation into seed dispersal by wind.

The largest seed in the world is 50 cm in diameter. It comes from the palm tree called Coco de Mer, found only in the Seychelles islands in the Indian Ocean. Another large seed is the coconut – it can be carried by the sea and germinate in a new place. Plants have developed many ingenious ways to be dispersed and to colonise new areas.

FIGURE 1.1.12a: Coco de Mer seed – the largest on the planet

Plants on the move? »

Plants cannot move. They colonise new areas by moving their **seeds** in a process called **dispersal**. Seeds can be dispersed by:

- wind
- water
- exploding pods that release seeds on touch or with moisture
- being carried inside animals that eat the fruit
- hooking onto the fur or skin of passing animals.

1. Look at Figure 1.1.12b, and identify how each of these seeds is dispersed.

2. Give reasons for all of your answers to question 1.

Ways of travelling »»

Seeds dispersed by wind have many shapes and sizes. The dandelion has parachute-like seeds, and the sycamore has seeds like helicopters. Peas and pansies have pods that explode when they have dried out or are touched by an animal, causing the seeds to fly out. Some plants produce fruits that animals eat but cannot digest. These pass through the animals, allowing the seed to **germinate** in another place using nutrients from the animals' dung. Burdock seeds have tiny hooks that catch on the fur of passing animals.

FIGURE 1.1.12b: Why have plants developed such variety in types of seed?

Seeds need to be dispersed as far away from the parent plant as possible, where there could be more light, nutrients and water – thereby increasing the chance of successful growth. Seeds are packed with nutrients to help the germinating plant to grow. Smaller seeds have fewer nutrients but may travel further. Larger seeds have bigger stores of food and can last much longer.

sycamore seeds

burdock seed

Alsomitra vine seed

FIGURE 1.1.12c: How are these seeds dispersed?

3. Why are the seeds from trees in forests most likely to be dispersed by the wind?

4. What are the advantages and disadvantages of a seed growing near the parent plant?

Surveying and sampling seeds »»

Botanists carry out surveys to try to find out how seeds are dispersed and how successful different plants are at germinating the seeds they make. They might do this by sampling many plants of the same species in a particular habitat. First they count the number of seeds made. Then, after the seeds have dispersed, they sample the habitat again to make an estimate of the number of seedlings that have germinated. By estimating the percentage of seedlings germinated compared to seeds made originally, they can judge how successful the seed dispersal mechanism is.

5. What is the **independent variable** in this survey?

6. What factors need to be controlled in such a survey?

7. How would you ensure the evidence collected was **reliable**?

8. Why might it be important to find out how successful plants are at dispersing and germinating seeds?

Did you know…?

The seeds of the *Alsomitra* vine tree were the inspiration in the development of the first gliders and airplanes. With a wing span of up to 13 cm, they are the largest wind-pollinated seeds in the world.

Key vocabulary

seed dispersal

germinate

independent variable

reliable

Understanding how fruits disperse seeds

We are learning how to:

- Describe how fruits are used in seed dispersal.
- Compare evidence about seed dispersal by wind and by fruit formation.
- Use data to evaluate different seed dispersal mechanisms.

Without animals to disperse their seeds, some plants would become extinct. The seeds of the *Astrocaryum* palm used to be dispersed by dinosaurs. Now, small rodents called agoutis disperse the seeds. Agoutis steal each other's seeds, increasing the distance of dispersal.

Plants exploiting animals

Plants use **fruits** to disperse seeds. A fruit is the **ovary** of a plant after fertilisation. The fruit is a nutritious treat surrounding the seed, mainly made of sugars and tasty nutrients to attract animals. Examples include nuts, tomatoes and cucumbers. The seed cannot be digested, so passes through the intestines and out with the faeces. Some seeds, such as mango seeds, are too large to be eaten. When they land in soil, they can germinate to make new plants.

1. What is a 'fruit'?

2. Which of the objects in Figure 1.1.13a is not a fruit?

3. What is the main advantage of fruits dispersing seeds?

FIGURE 1.1.13a: Where are the seeds in these plant products?

Why seeds are dispersed

In producing fruit a plant uses energy, which is transferred to the animals that eat the fruit. The advantage for the plant is that it does not need to produce as many seeds, and most are carried away from the parent plant and land in nutritious soil.

Plants that use wind to disperse seeds usually produce thousands of much smaller seeds, to increase the chance of successful germination. Their small size and aerodynamic features allow them to be dispersed over much larger areas, but with no guarantee of landing in nutritious soil.

The rubber plant and witch hazel have exploding pods that burst open when the seeds are ripe. This mechanism guarantees dispersal, though rarely very far from the parent plant, increasing **competition** and reducing the chance of successful germination. Coconuts are large and buoyant and so can be tranported by seas. They are packed with nutrition so the seed can survive a long time.

4. Which type of dispersal mechanism requires more seeds? Why is this?

FIGURE 1.1.13b: How are these seeds adapted for their dispersal method?

Methods of seed dispersal ⟩⟩⟩

TABLE 1.1.13: Different methods of seed dispersal

Name of plant	Type of dispersal mechanism	Approximate number of seeds made per plant	Average dispersal distance
ragwort	parachute	10 000	over 100 m
ash tree	helicopter	1000	over 100 m
Alsomitra vine tree	glider	40 000	1–2 km
witch hazel	exploding pod	100	10 m
pea	exploding pod	100	a few metres
blackcurrant	fruit	300	variable
melon	fruit	500	variable
coconut	water	50	hundreds of miles

Table 1.1.13 summarises the types of seed dispersal mechanisms used by a variety of plants.

5. If you were a plant, which dispersal mechanism would you choose and why?

6. What can you say about the different dispersal mechanisms from the data?

7. Show the data from the table in a graphical form. Choose a good way to represent the data so that mechanisms can be evaluated.

> **Did you know…?**
>
> Avocados are thought to be the most nutritious fruit with over 25 essential nutrients, including vitamin C, iron, magnesium and potassium. Eating plenty of different fruit can reduce the risk of cancer, heart disease, strokes and Alzheimer's disease.

Key vocabulary

fruit

ovary

competition

Understanding the male reproductive system

We are learning how to:

- Describe the structure and function of different parts of the male reproductive system.
- Compare plant and human male reproductive structures.
- Summarise the strengths and weaknesses of the human and plant male reproductive systems.

The human reproductive system is controlled by chemicals. In the male, one chemical is testosterone, which controls the growth and development of the organs and sperm cells. Sperm cells take four to six weeks to mature, and live for about 36 hours once released inside the female.

Male and female

The purpose of a **reproductive system** is to produce offspring and so keep the species alive. In some species, such as plants, the male and female organs are on the same organism. Most vertebrates have separate male and female organisms, with specially adapted reproductive systems. The purpose of the human male reproductive system is to make millions of male sex cells (sperm) and to transport them inside the female to fertilise an egg cell and so produce a baby.

1. Name the male sex cells and female sex cells in humans.

2. What is the purpose of the male reproductive system?

Naming the parts

The **testes** are two organs where human sperm cells are made. They are protected inside the **scrotal sac**. A tube called the **sperm duct** carries the sperm to a large organ called the prostate gland. Here, a liquid called **semen** is produced and mixed with the sperm cells, to supply them with nutrients for their long journey. They leave the male through a tube called the **urethra**, inside the organ known as the **penis**. This occurs during the act of sexual intercourse.

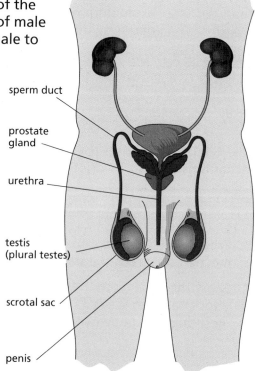

sperm duct

prostate gland

urethra

testis (plural testes)

scrotal sac

penis

FIGURE 1.1.14a: The male reproductive system

3. List one cell and two organs in the male reproductive system.

4. Draw the journey of a sperm cell, labelling the parts of the male reproductive system that it passes through.

Transfer of the male sex cell

The human male reproductive system has the same purpose as the stamen in a flower. The anther makes pollen (the male sex cells of the plant), just as the testes make sperm cells. The anther releases the pollen to be transported to the stigmas of other plants using external influences such as insects or the wind.

Humans carry out internal fertilisation. In sexual intercourse, the penis is inserted inside the vagina – its movement stimulates the release of sperm from the testes. In this way, sperm are guaranteed to be placed directly inside the female. Both the anther and the testes produce millions of male sex cells to maximise the likelihood of successful fertilisation. However, plants produce pollen only when the stigmas are likely to be ready for fertilisation.

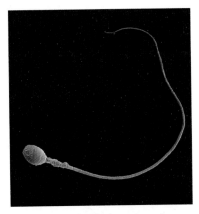

FIGURE 1.1.14b: The human male sex cell is adapted to carry out its job. How is it different from a pollen cell?

5. Which parts make the male sex cells in plants and in humans?

6. What additional features does a plant reproductive system have that a human reproductive system does not have? Why does it need them?

7. What advantage(s) does the plant male reproductive system have over the human male reproductive system?

8. Which system do you think is more effective – the stamen or the human male reproductive system? Give reasons for your answer.

Did you know…?

A human sperm is the smallest cell in the body. 5000 sperm cells would fit into one millimetre. The egg cell is the largest – about the size of a full stop.

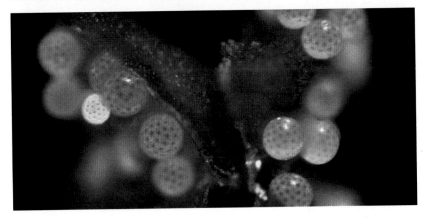

FIGURE 1.1.14c: Pollen on the stamen of a flower. Compare this with the human male reproductive system.

Key vocabulary

reproductive system

testes

scrotal sac

sperm duct

semen

urethra

penis

Understanding the female reproductive system and fertility

We are learning how to:

- Describe the structures and functions of different parts of the female reproductive system.
- Explain the process of fertilisation.
- Explain problems of infertility and how they might be treated.

The human female reproductive system receives sperm and enables the fertilised egg to develop until it is ready to be born. The uterus, or womb, is where the foetus grows and develops. The uterus increases to up to 20 times its original size during pregnancy.

The functions of female organs

The human female reproductive system has two main purposes – to produce egg cells that may be fertilised by the male sperm, and to provide an environment for the growing foetus.

The main female organs are the **vagina**, **cervix**, **uterus**, **oviduct** and **ovary**. Table 1.1.15 summarises the structure and function of each of these.

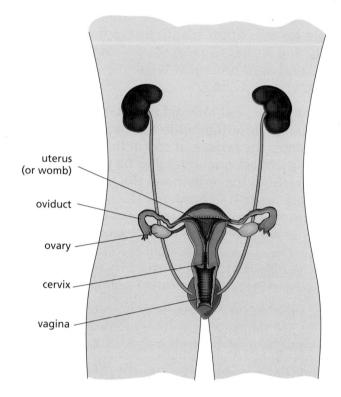

uterus (or womb)

oviduct

ovary

cervix

vagina

FIGURE 1.1.15a: The female reproductive system

TABLE 1.1.15: Female reproductive organs

Vagina	Muscular tube, 8 to 12 cm long, that extends up to the uterus and can stretch to allow a baby to pass
Cervix	Narrow opening from the vagina to the uterus with thick walls – can extend wide enough to allow a baby to pass
Uterus or womb	Pear-shaped cavity with thick muscular walls – where fertilisation occurs and the developing baby grows
Oviduct (Fallopian tube)	The tube that carries the egg from the ovary to the uterus
Ovary	Where eggs cells are made and then released into the oviduct

1. Where are female sex cells made?

2. Why do you think the uterus has muscular walls?

One egg matures each month in the ovary and is released into the oviduct – this process is called **ovulation**. The lining of the oviduct contains specialised cells with tiny hairs that beat causing the egg to move down to the uterus, where for up to 24 hours it may be fertilised by a sperm cell. Only one sperm penetrates the egg cell, losing its tail as it does so. The nucleus of the sperm fuses with the nucleus of the egg, combining the genetic material of both. The fertilised egg is the start of a new life.

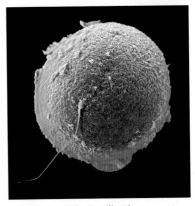

FIGURE 1.1.15b: Fertilisation occurs when one sperm cell penetrates the egg cell and their nuclei fuse.

3. Why does the egg need to move from the ovary to the uterus?

4. Why do you think the egg cell is so much bigger than the sperm cell?

5. Explain the difference between ovulation and fertilisation.

Infertility >>>>

Most women below the age of 36 have little trouble in having babies. However, **infertility** affects about 3.5 million women in the UK. There are a number of causes:

- External factors – such as excessive alcohol, drugs, long-term smoking, stress and sexually transmitted diseases.

- Problems with ovulation – the release of eggs is controlled by chemicals called hormones; an imbalance may result in eggs not being made or released.

- Endometriosis – cells from the lining of the oviduct may start to grow around the ovary and cause cysts to appear, making it hard for the eggs to be released.

- Blockages in the oviduct – these can prevent an egg from reaching the uterus and becoming fertilised.

In men, infertility may be caused by a low number of healthy sperm or sperm that can't swim well because of disease.

6. For each of the female infertility problems, suggest a possible solution.

Did you know…?

The ovaries of new-born girls have about 600 000 immature eggs. However, an adult woman is capable of giving birth to a maximum of 35 babies.

Key vocabulary

vagina

cervix

uterus

oviduct

ovary

ovulation

infertility

Learning about changes in puberty

We are learning how to:

- Recognise changes in male and female bodies during puberty.
- Describe the process of menstruation.
- Explain how some problems with menstruation occur.

Puberty refers to the period when physical changes occur that enable a person to reproduce. It can start from any age between 8 and 16 years. All the changes are controlled by chemicals called hormones, which work together.

Changes during puberty

During **puberty** in girls:

- the hips widen, preparing for childbirth
- there is a height spurt
- breasts become bigger to prepare for breastfeeding
- **menstruation (periods)** start
- hair develops in the armpits and around the reproductive organs.

During puberty in boys:

- the shoulders broaden to give a strong appearance
- the voice deepens to attract females
- there is a height spurt
- the penis and testes grow
- sperm is produced and released during 'wet dreams' to prepare for intercourse
- hair develops in the armpits and around the reproductive organs.

1. What changes in puberty occur in *both* boys and girls?

When the changes happen

The changes that occur during puberty, and the ages between which they usually occur, are shown in Table 1.1.16.

Event in puberty	Age range for start of event (years)
height spurt in girls	8.5 – 14.0
development of breasts	8.0 – 13.0
first menstruation period	10.5 – 15.5
height spurt in boys	10.5 – 16.5
growth of penis	10.5 – 14.5
growth of testes	9.5 – 13.5
growth of voice box (larynx)	10.5 – 14.0

FIGURE 1.1.16a: Individuals experience changes at different ages.

TABLE 1.1.16: Age ranges for the start and end of changes in puberty

2. Look at the data in Table 1.1.16. What are the youngest ages at which a girl and a boy can start puberty?

3. Is there any order in which particular events occur in puberty? Give evidence for your answer.

Menstruation »»»

Menstruation occurs in a cycle lasting about 28 days and is controlled by hormones. Some women experience problems:

- Amenorrhea (absence of periods) is caused by hormonal problems, defects in the ovary, stress or anorexia.

- Menorrhagia (excessively heavy bleeding) is caused by hormonal imbalances or infection in the uterus.

- Dysmenorrhea (period pain) is also caused by hormone imbalances.

> **Did you know…?**
>
> Apart from humans, only primates, elephant shrews and some bats undergo menstruation. In other mammals, the lining is reabsorbed by the body so that nutrients are not lost.

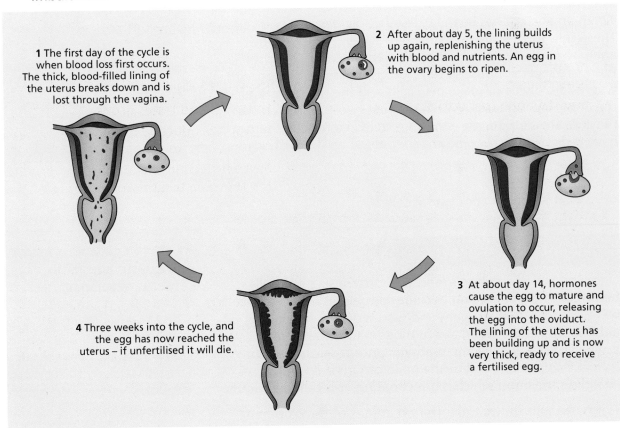

1 The first day of the cycle is when blood loss first occurs. The thick, blood-filled lining of the uterus breaks down and is lost through the vagina.

2 After about day 5, the lining builds up again, replenishing the uterus with blood and nutrients. An egg in the ovary begins to ripen.

3 At about day 14, hormones cause the egg to mature and ovulation to occur, releasing the egg into the oviduct. The lining of the uterus has been building up and is now very thick, ready to receive a fertilised egg.

4 Three weeks into the cycle, and the egg has now reached the uterus – if unfertilised it will die.

FIGURE 1.1.16b: The menstrual cycle

4. What part do hormones play in normal and abnormal periods?

5. Look at Figure 1.1.16b. Explain the problems that may occur at different stages of the cycle.

Key vocabulary

puberty

menstruation

period

Learning how a foetus develops

We are learning how to:

- Recognise the process of growth.
- Use data to show how the embryo grows during gestation.
- Compare and contrast the pregnant uterus with the non-pregnant uterus.

A human foetus takes 38 weeks to grow from one fertilised cell into a complete baby ready to be born. Dogs take just two months, whereas elephants take up to two years. The mother provides the developing foetus with all the nutrients and oxygen it needs, as well as removing all waste products.

Cell division

When an egg cell has been fertilised, it divides into two cells. These cells further divide to make four cells, which divide again to make eight cells. This **cell division** continues until there are several thousand cells. This is the process of growth, where cells divide to make new cells and the overall size of the organism increases. Within the first two to three weeks the cells are all the same – they are called **stem cells**. Stem cells have the ability to become any specialised cell in the body.

1. What is 'growth'?

2. What is special about stem cells?

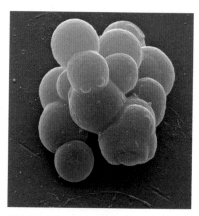

FIGURE 1.1.17a: Stem cells

Development of the foetus

Once the ball of stem cells reaches a certain size, the cells begin to differentiate and become specialised cells. Some cells will develop into the organs and tissues of the developing baby. At this stage when the cells begin to differentiate, the ball of cells is called an **embryo**. Once it reaches about 8 weeks old, when most of the main organs are formed, including the heart which is now beating, it is called a **foetus**.

Figure 1.1.17b shows the different stages of development of a human foetus. Ultrasound is used to make images of the foetus at different stages to monitor its development and identify any problems. The height of the foetus can be measured using these images.

3. When is the fastest period of growth of the developing foetus? Explain your answer.

Did you know...?

The taste buds of a foetus develop at 14 weeks; it can hear at 24 weeks and track objects with its eyes at 31 weeks. At 28 weeks, a foetus is likely to survive if born.

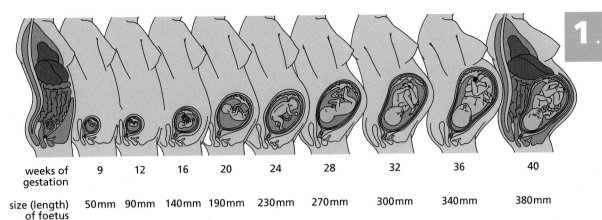

weeks of gestation	9	12	16	20	24	28	32	36	40
size (length) of foetus	50mm	90mm	140mm	190mm	230mm	270mm	300mm	340mm	380mm

FIGURE 1.1.17b: Foetuses at different stages of development

Supporting structures ⟩⟩⟩

During pregnancy, other cells from the original ball of cells will become structures that connect with the mother – the **placenta**, amnion, amniotic fluid and **umbilical cord**. These structures are shown in Figure 1.1.17c.

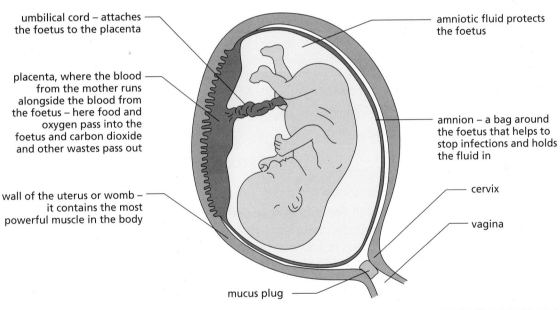

umbilical cord – attaches the foetus to the placenta

placenta, where the blood from the mother runs alongside the blood from the foetus – here food and oxygen pass into the foetus and carbon dioxide and other wastes pass out

wall of the uterus or womb – it contains the most powerful muscle in the body

amniotic fluid protects the foetus

amnion – a bag around the foetus that helps to stop infections and holds the fluid in

cervix

vagina

mucus plug

FIGURE 1.1.17c: The developing foetus in the uterus

4. Why does a foetus need the placenta?

5. Why is it important for the baby to be surrounded by fluid?

6. Summarise the different ways in which a pregnant uterus is different from a normal uterus.

Key vocabulary

cell division

stem cells

embryo

foetus

placenta

umbilical cord

Understanding factors affecting a developing foetus

We are learning how to:

- Describe the effects of different factors on a developing foetus.
- Evaluate the strength of data.

A foetus can't take in its own food or oxygen and relies on the mother to supply it with essential chemicals and nutrients. The placenta allows substances to pass from mother to baby.

The role of the placenta

The placenta allows oxygen, glucose, digested proteins and fats, vitamins and minerals to enter the foetus – it also removes carbon dioxide and waste products, such as urea. Harmful substances can also cross the placenta including alcohol, nicotine, carbon monoxide, cocaine, insecticides, lead and mercury.

1. How might the harmful substances come to be at the placenta?

2. What would happen to a foetus without the placenta? Explain your answer.

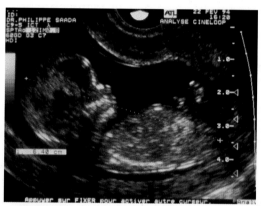

FIGURE 1.1.18a: An ultrasound scan of a foetus enables its development to be checked.

Effects of substances on the foetus

Scientific studies have established how different substances affect a developing foetus. Foetal size and movements can be tracked and the heartbeat measured. Tests have found out that some substances affect the foetus – see Table 1.1.18.

TABLE 1.1.18: Substances that affect a developing foetus

Alcohol	Higher rate of stillbirth, lower birth weight, lower IQ; baby slower to move and think, more likely to be dependent on alcohol in adulthood.
Smoking – nicotine and carbon monoxide	Much higher risk of stillbirth, **premature** delivery and low birth weight resulting in poor development; greater likelihood of developing asthma.
Drugs – marijuana, cocaine	Higher rate of stillbirth, premature birth, low birth weight, learning difficulties and likely addiction to the drug.
Nutrition – folic acid	Good for the development of the brain and spinal cord; supplements should be taken as soon as pregnancy is recognised.

3. What are the common factors that badly affect the development of a foetus?

4. What advice can you give to pregnant mothers to help them have a healthy baby?

5. How confident can you be about the evidence produced by ultrasound scans? Explain your answer.

Validity and reliability in research

Researchers need to ensure that their investigations produce **valid** and **reliable** evidence. 'Valid' means that the evidence collected answers the question being investigated. It must take account of all possible variables. The evidence should also be reliable. This can be done through repeat readings or, in the case of a survey, using a large **sample size**.

> ### Did you know…?
>
> A woman may not realise she is pregnant until about 8 weeks after conception. The embryo's brain starts to develop after just 2 to 3 weeks and is highly influenced by chemicals coming through the placenta.

6. Comment on the validity and reliability of the following studies:

   a) The first research on the effects of alcohol was conducted on 127 babies born to alcoholic mothers in France in 1968. The babies were found to have lower birth weights and lower intelligence.

   b) In a study on the effect of smoking, the ultrasound scans of 65 mothers who smoked were compared with the scans of 36 mothers who were non-smokers.

   c) In a study on the use of folic acid, the mothers of 85 per cent of Norwegian children born between 2002 and 2008 completed a questionnaire. Researchers found that 0.1 per cent of mothers who took folic acid had autistic children, compared to 0.21 per cent who did not take folic acid.

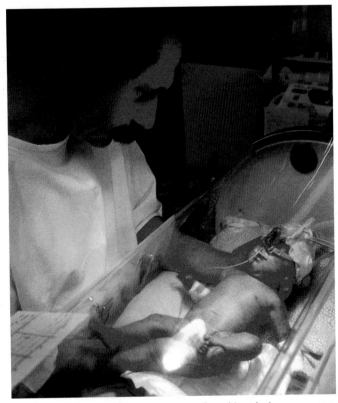

FIGURE 1.1.18b: Possible consequence of smoking during pregnancy

Key vocabulary

premature

valid

reliable

sample size

Checking your progress

To make good progress in understanding science you need to focus on these ideas and skills.

Recognise and label basic and specialised animal cells and plant cells; use a microscope to make observations.

Describe the functions of the nucleus, cell membrane, mitochondria, cytoplasm, cell wall, vacuole and chloroplast.

Compare and contrast the similarities and differences between specialised animal cells and plant cells.

Describe unicellular organisms – including yeast, bacteria, euglena, paramecium and amoeba – as being either prokaryotes or eukaryotes.

Describe the function of specialised parts of different unicellular organisms.

Explain how different structures help organisms to survive.

Recognise the role of diffusion in living organisms.

Describe the process of diffusion, and name the materials needed by the cell and those removed from the cell.

Explain the factors that affect diffusion.

Put the terms cell, tissue, organ and organ system in order of hierarchy, naming some common tissues, organs and organ systems in humans.

Explain the terms cell, tissue, organ and organ system and the function of all the main organ systems in the body.

Describe some benefits and disadvantages of multicellular organisms compared to single-celled organisms.

Describe the role of different parts of a flowering plant in reproduction.

Explain the differences in wind-pollinated and insect-pollinated plants.

Discuss the strengths and weaknesses of wind-pollinated and insect-pollinated plants.

☐ Recognise different seed-dispersal methods and relate these to the structures of the seeds.

☐ Identify key variables that need to be controlled when investigating the effect of seed design on seed dispersal.

☐ Explain the advantages and disadvantages of different seed-dispersal mechanisms.

☐ Name the main parts of the male and female human reproductive systems.

☐ Describe the structures and functions of the main parts of the male and female human reproductive systems; describe how fertility problems may arise.

☐ Explain how the male and female reproductive structures are designed for fertilisation; describe methods to combat infertility.

☐ Recognise changes that occur during adolescence.

☐ Describe how the menstruation cycle works.

☐ Explain how and why some problems occur with menstruation.

☐ Identify substances passed on from a mother that will either help or harm her developing foetus.

☐ Describe the structures and functions of different parts of a pregnant uterus, describing how substances pass into and from a developing foetus.

☐ Explain how a pregnant uterus is different from a normal uterus, including the impact of different substances on the health and development of a foetus.

Questions

Questions 1–7

See how well you have understood the ideas in the chapter.

1. Which of the following is a unicellular organism? [1]

    a) nerve cell   **b)** cytoplasm   **c)** amoeba   **d)** flowering plant

2. Where in the cell would the most diffusion take place? [1]

    a) nucleus   **b)** cell membrane   **c)** chloroplast   **d)** cell wall

3. Which structure is not directly linked to fertilisation? [1]

    a) egg cell   **b)** ovary   **c)** stigma   **d)** pollen grain

4. Which of the following will not pass from a mother to her developing foetus across the placenta? [1]

    a) carbon dioxide   **b)** carbon monoxide   **c)** alcohol   **d)** glucose

5. Using an example, describe the theory of 'spontaneous generation'. [2]

6. Describe the events after pollination that lead to fertilisation. [2]

7. Outline what happens in the menstruation cycle. [4]

Questions 8–14

See how well you can apply the ideas in this chapter to new situations.

8. Some plants live in conditions of low light on the floor of thick forest. Which of the following features are likely to help them to survive? [1]

    a) They will have brightly coloured petals.

    b) Their leaves will be dark green, packed with more chloroplasts than ordinary leaves.

    c) They will have fewer root hair cells.

    d) Their seeds will have parachutes so they can be blown by the wind.

9. Cells in the lining of the human lung need to transfer oxygen quickly from the lungs to the blood. How are the cells likely to be adapted to carry out their job? [1]

    a) They will contain chloroplasts to collect sunlight.

    b) They will contain cilia to remove bacteria.

    c) They will have a thin cell membrane and lots of mitochondria.

    d) They will have a large surface area and a thin cell membrane.

10. Insect populations in towns are declining. What can be done to increase these populations? [1]

    a) Grow a greater variety of wild flowers.   **b)** Use more pesticides.

    c) Grow more crops for food.   **d)** Build more roads and buildings.

11. Seeds are dispersed by a variety of mechanisms. Some are shown in Figure 1.1.20a. Which type of seed is likely to disperse the furthest if the plant was growing on an island? [1]

	Type of seed
a)	Avocado stone
b)	Coconut
c)	Dandelion head
d)	Burdock seed

 Burdock seed
 Avocado stone
 Dandelion head
 Coconut

FIGURE 1.1.20a

12. Some scientists discover a new unicellular organism. What features would enable them to classify it as algae? [2]

13. A sixteen-year-old girl has not yet begun to menstruate. What two reasons could there be? [2]

14. Bees in Australia are not affected by colony collapse disorder. Explain why this might be so. [4]

Questions 15–16

See how well you can understand and explain new ideas and evidence.

15. Figure 1.1.20b shows the percentage of motile (moving) sperm in men according to whether they are smokers and/or fertile. What information would you want to know about this study before you could trust the data? [2]

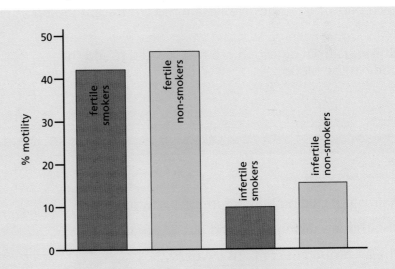

FIGURE 1.1.20b: How smoking affects motility of sperm
(Source: Bethany R. Brookshire, Ph.D scientopia.org/blogs/scicurious/)

16. Sketch a graph to show how the weights of foetuses from smoking mothers compare to those from non-smoking mothers. Give reasons for the differences. [4]

Eating, Drinking and Breathing

Ideas you have met before

Diet and nutrition

Animals cannot make their own food and must eat plants or other animals for energy.

Humans must eat a balanced diet containing the correct types of food to stay healthy.

Digestion

We have different types of teeth and each type has a different role in breaking down food.

Several parts of the body help us to digest food – such as teeth, stomach and intestines.

Each part of our digestive system has a different job to do.

Nutrients from digestion are transported round the body in the blood.

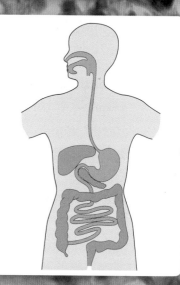

Breathing and gas exchange

Animals, including humans, need air to survive.

Breathing is taking air in and out of our lungs.

The air around us contains oxygen.

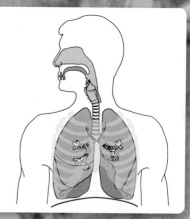

A healthy diet

- What each kind of food in a healthy diet does for us.
- How much energy we need and how much we get from the different foods we eat.
- How eating too much or too little can affect us.

The digestive system

- The role of each part of the digestive system.
- How each organ in the digestive system is well designed to do its job.
- What happens to food molecules after digestion.
- What enzymes are and how they help with digestion.
- How bacteria help us to digest food.

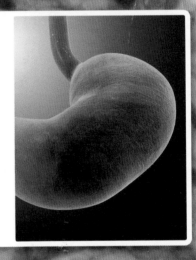

The breathing system

- How humans breathe.
- How parts of our breathing system are well designed to get gases in and out of our bodies.
- How oxygen from breathing is used in the body.
- How disease and lifestyle can affect our breathing system.

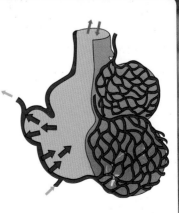

Exploring a healthy diet

We are learning how to:

- Describe the components of a healthy diet.
- Examine the importance of each component of a healthy diet.
- Evaluate the quality of evidence contained in advertising about a healthy diet.

Some adverts try to promote food products as being part of a healthy diet. What is a healthy diet and should you believe all the adverts?

Food groups

There are seven **food groups**, shown in Table 1.2.2. These provide the **nutrients** you need to live, grow and use in processes in the body.

TABLE 1.2.2: The seven food groups

Food group	Uses in the body
carbohydrates	Two types: starches and sugars. They provide energy – an excess causes weight increase.
protein	Important for growth and repair.
fats and oils	Stored as a reserve energy supply. A layer under the skin provides insulation against cold. An excess causes weight gain and can lead to other health issues.
minerals	Tiny amounts are needed – e.g. iron for red blood cells and calcium for bones.
vitamins	Small amounts are needed – e.g. vitamin C for repair of the skin and vitamin D for taking up calcium.
fibre	Needed to keep the large intestine working well.
water	Needed to stop a person becoming dehydrated.

starchy foods – bread, potatoes, pasta

fruits and vegetables

milk and dairy products

foods high in fat and sugar

protein foods – meat, fish, eggs

FIGURE 1.2.2a: You can help to ensure that you have some of each food group in your diet by varying the foods that you eat.

1. Name the main food group we get from fish.

2. Give two foods that contain carbohydrates.

3. Name two food groups we get from cheese.

A balanced diet »»

Each of the food groups has a different role to play in the **balanced diet** that our bodies need.

In a typical Western diet, there is often too much sugar and fat and not enough fibre. Fibre is an unusual food group because we do not digest it. Instead, it adds bulk to food and helps to move it through the intestines. Fibre is mainly cellulose, from plant cell walls. A lack of fibre can cause constipation, where faeces become difficult to pass.

A Mediterranean diet consists of lots of fruit, vegetables and grains, and fats from healthy sources, such as olive oil. It is believed by some that it is linked to good health and a lowered risk of heart disease.

4. We are sometimes told that fat in our diet is bad for us. Explain why this can be both true *and* false.

5. Explain why constipation is less common in Mediterranean countries than in most Western countries.

Healthy eating plans »»»

We need different amounts of each of the food groups in our diet – it would not be sensible, for example, to eat the same amounts of protein and fats. An 'eatwell plate' like that shown in Figure 1.2.2a can help us make sure that we are eating the food groups in the correct proportions.

Advice on healthy eating can be found in magazines, on TV and on the internet. It sometimes comes from companies who want to sell eating plans and diet supplements. Only some of the advice can be trusted.

6. List the food types in order of the relative proportions we should include in our diet, as recommended by the 'eatwell' plate in Figure 1.2.2a.

7. Describe the evidence (if any) given in the adverts in Figure 1.2.2b and 1.2.2c for these diets improving health.

8. Should you believe adverts? What might you want to know before deciding whether or not to trust them?

A study of 1.5 million adults has shown that a Mediterranean diet leads to reduced risk of heart disease and cancer.

FIGURE 1.2.2b: Would this advert make you include olive oil in your diet?

A vegetarian diet reduces the risk of bowel cancer linked to constipation.

FIGURE 1.2.2c: Would this advert make you change what you eat?

Key vocabulary

food group

nutrient

balanced diet

Testing foods

We are learning how to:

- Test foods for starch, sugars, protein and fat.
- Predict the results of food tests for a range of foods.
- Evaluate the risks involved in carrying out food tests.

Food manufacturers must label all their foods to show exactly what they contain. They employ scientists to find out this information.

Importance of food labels

In the UK, the government is in charge of ensuring that all packaged foods are labelled to show what they contain and the different food groups in the food. Accurate labelling allows people to make healthy choices about what they are eating and allows people with allergies to avoid certain foods.

1. Who is responsible for accurate labelling of foods in the UK?

2. List some examples of the type of information included on food labels.

3. Suggest two reasons why foods must be correctly labelled.

FIGURE 1.2.3a: Why is it important that the information on labels is accurate?

Testing for different food groups

We can find out which food groups a food contains by testing it using chemicals. Scientists carry out the tests to allow foods to be labelled correctly. The tests include:

- **Starch**: a colour change with iodine solution from orange to blue/black.

- **Sugar**: a colour change with Benedict's solution from blue to orange.

- **Protein**: a colour change with Biuret solution from blue to mauve.

- **Fat**: appearance of a white milk-like solution when ethanol is added.

Table 1.2.3 shows the results from tests carried out by scientists working for a food manufacturer.

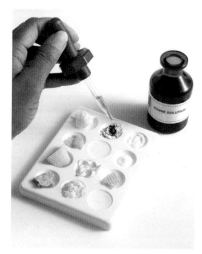

FIGURE 1.2.3b: Testing food for starch

TABLE 1.2.3: What do the results show about each of the foods?

Food	Observations				Conclusion
	Test for starch	Test for sugar	Test for protein	Test for fats	
pasta	positive	negative	negative	negative	pasta contains starch
chicken	negative	negative	positive	positive	chicken contains protein
chickpeas	positive	negative	positive	negative	chickpeas contain protein and starch

4. The conclusion made in Table 1.2.3 about the food groups that chicken contains is incomplete. Name the other food group that it contains.

5. Describe the observations that the scientists would have made when testing chickpeas.

6. Predict the observations that the scientists would make if they tested milk.

Minimising risks

Most practical procedures that scientists carry out have some **risk**. For example, they may use potentially harmful chemicals, boiling water or a naked flame. It is the scientist's job to identify risks, decide whether they can be minimised and then to decide whether it is worth taking that risk. Students working as scientists in school need to do this, too.

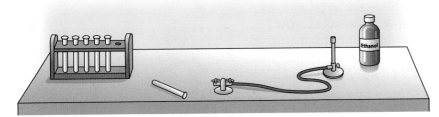

FIGURE 1.2.3c: Are there any risks associated with the methods used to test foods?

7. Identify any risks associated with the practical methods for testing foods in the school laboratory.

8. Explain how students can minimise the risks.

9. Imagine that some students do not have a water bath and are considering heating an ethanol solution using a Bunsen burner. Evaluate the risk associated with this and suggest whether or not they should take it.

Key vocabulary

starch

sugar

protein

fat

risk

Comparing energy needs

We are learning how to:

- Describe how we use energy from food.
- Compare the energy requirements of people of different ages and with different lifestyles.
- Analyse numerical data about energy contents of different foods.

If you spend a day sitting in front of a computer you will use less energy than if you spend the day competing at a sporting event. The amount of energy that we need varies with age and level of activity. However, we need energy even when we are still.

Using energy

You get your **energy** from food. Energy is transferred from food into your body by a process called **respiration**.

You need energy to:

- grow
- repair
- move
- keep warm.

FIGURE 1.2.4a: Is this person using energy?

When you move, muscles contract. Muscles need energy to contract. Even when you are still and sleeping, your body needs energy. Your heart is a muscle that continually contracts, muscles in your chest move as you breathe and muscles in arteries help your blood to flow.

1. Where does your energy come from?
2. List four reasons why you need energy.
3. Explain why you use energy even when you are sleeping.

Energy needs vary

The amount of energy in foods is measured in **kilojoules (kJ)**. This can be found listed on some food labels. The amount of energy we need per day varies, depending on age, gender and level of activity.

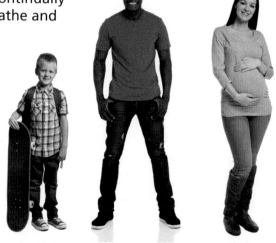

| 6500 kJ | 11 500 kJ | 10 000 kJ |

FIGURE 1.2.4b: Why do these people need different daily amounts of energy?

4. Suggest why there might be a difference between the daily energy requirements of an average adult woman and a pregnant woman.

5. Suggest why males generally need more energy per day than females.

6. **a)** Predict how the energy requirements of an elderly man compare with that of an average adult man. Explain your answer.

   b) How do you think the energy requirements of an elderly man compare with those of an elderly woman? Explain your answer.

Keeping track of the energy in our food

Some food labels show the amount of energy that the food contains. These labels help people to know the amount of energy they are taking in. However, we should also take into account other factors when choosing which foods to eat, such as the nutrients and the amounts of salt and fat they contain.

7. Looking at just the energy counts shown in Figure 1.2.4c, suggest why some people would find it difficult to choose between having butter or jam for breakfast, or between having chocolate or a banana for a snack.

8. Think back to what you have learned about nutrients in food. Which of the choices in question 7 would be the healthier? Explain why.

9. Calculate how many bananas an average adult male should eat to obtain the energy he requires if he ate nothing else for the whole day. What would your advice be to this man about eating in this way?

Did you know…?

Three iced ring doughnuts contain approximately the amount of energy an 8-year-old boy needs during one day. It would take over 90 minutes of cycling to use up this energy.

Nutrition Facts	
Butter	
Amount per serving	
Energy	255 kJ

Nutrition Facts	
Apricot jam	
Amount per serving	
Energy	250 kJ

Nutrition Facts	
Chocolate	
Amount per serving	
Energy	548 kJ

Nutrition Facts	
Banana	
Amount per serving	
Energy	557 kJ

FIGURE 1.2.4c: Some food labels tell us how much energy is in each portion of food.

Key vocabulary

energy

respiration

kilojoule (kJ)

Exploring obesity and starvation

We are learning how to:

- Describe the physical effects of eating too much and eating too little.
- Explain the physical effects of obesity and starvation.
- Compare how deaths from starvation and obesity have changed over time.

We can all eat too much or too little at times. For example, you may eat too much during holidays or eat less than usual if you are unwell. However, when this is prolonged, you can cause serious damage to your health.

The effects of eating too much or too little

Body mass index (BMI) is a measure of body fat based on the height and weight of an individual. People with **obesity** have a very high BMI. Obesity can cause serious physical problems such as pain in the joints, heart disease, high blood pressure and difficulties with breathing.

Starvation is caused by not taking in enough energy and nutrients over a prolonged period. It is the most serious form of **malnutrition** and can cause physical problems such as severe weight loss, muscle loss, dry skin and hair, infertility and fatigue. Both obesity and starvation can cause death.

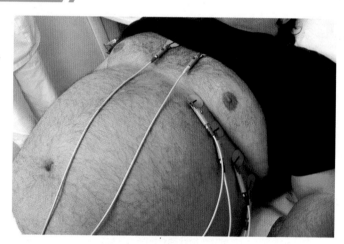

FIGURE 1.2.5a An obese person is at greater risk of health problems.

1. What is meant by 'BMI'?

2. What causes starvation?

3. Draw a table to summarise the physical effects of obesity and starvation.

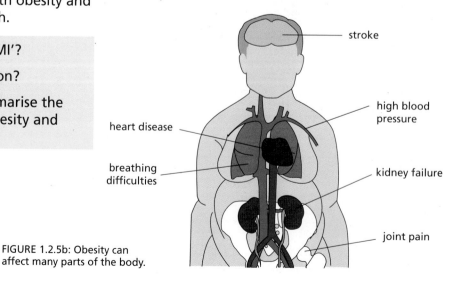

stroke

high blood pressure

heart disease

breathing difficulties

kidney failure

joint pain

FIGURE 1.2.5b: Obesity can affect many parts of the body.

Each of the physical effects of obesity and starvation can be explained. With obesity, extra fatty tissue is stored in the body. The body then struggles to carry the increased mass. This can put strain on the joints. The heart is also put under strain because it has to pump blood to the extra fatty tissue.

With starvation, the body has insufficient energy and nutrients – such as vitamins, minerals and proteins – to keep the person healthy. The body uses up fat stores and the person loses excessive weight. Muscles may also be used as an energy source.

4. Explain how both obesity and starvation can lead to fatigue.

5. Write explanations for joint problems in obesity and for poor condition of skin and hair in starvation.

Measuring deaths caused by obesity and starvation ⟩⟩⟩

Scientists try to monitor the causes of death of people around the world. The graph in Figure 1.2.5d shows an estimate of the number of deaths caused by obesity and starvation in 1990, 2000 and 2010.

6. Describe the trend in the number of deaths between 1990 and 2010 caused by:

   a) obesity

   b) starvation

7. Compare the number of deaths from obesity and starvation in 2010.

8. Suggest why it is difficult to gain accurate numbers for the deaths caused by obesity and starvation.

FIGURE 1.2.5c: This child is malnourished and starving. The swollen abdomen indicates a lack of protein.

Did you know...?

Obesity has many causes, including overeating, a genetic tendency to gain weight, psychological issues and other medical issues such as thyroid problems.

Key vocabulary

obesity

starvation

malnutrition

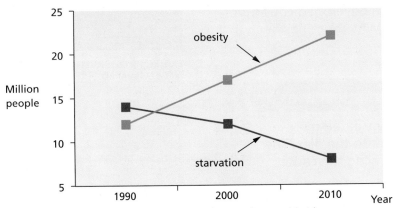

FIGURE 1.2.5d: Deaths caused by obesity and starvation worldwide.

Understanding deficiency diseases

We are learning how to:

- Identify the causes and effects of some deficiencies in the diet.
- Suggest which foods could prevent well-known deficiencies.
- Plan ways of communicating ideas about preventing deficiency diseases.

No matter how much you eat, you can still be deficient in some nutrients if your diet is not balanced. Poor nutrition, or malnutrition, can cause severe disease and even death. In the past the causes of many of these deficiencies have been a mystery.

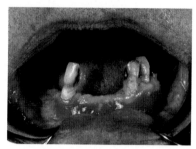

FIGURE 1.2.6a: Scurvy is caused by a lack of vitamin C.

Diseases caused by deficiencies

Many **deficiency diseases** are caused by a lack of certain **vitamins**. However, a person can also suffer from a lack of other food groups such as minerals or protein.

The symptoms of **scurvy** are fatigue, bleeding gums, loss of teeth, fever and death. It is caused by a deficiency of vitamin C. In the past it caused the death of sailors whose diet consisted of dried meat and grains, but little fresh fruit.

Rickets causes muscles and bones to become soft. This can lead to permanent deformities in children. Rickets is caused by a lack of vitamin D, which is needed to absorb calcium into bones to strengthen them and allow growth. Sunlight is needed to allow your body to use vitamin D.

1. In the past, why did a lot of sailors suffer from scurvy?

2. Suggest why children need more calcium than adults.

3. In recent years, children living in developed countries have spent more time indoors. Why there has been a rise in rickets in recent years?

FIGURE 1.2.6b: Rickets is caused by a lack of vitamin D.

Preventing and treating deficiency diseases

Deficiency diseases are usually easily treated by re-introducing the missing nutrient into the diet. The problem in the past has been identifying the cause.

Scurvy is treated by eating foods containing vitamin C (or taking supplements). Some foods, such as milk formula and cereals, have vitamin D added to them to prevent rickets.

TABLE 1.2.6: Deficiency diseases

Disease	Deficiency causing the disease	Food to treat or prevent the disease
scurvy	vitamin C	limes
rickets	vitamin D	eggs
anaemia	iron	beans and pulses

4. What does the information in Table 1.2.6 tell you about the nutrient content of beans and pulses?

5. From Figure 1.2.6c, identify two foods to treat or prevent each of scurvy, rickets and anaemia.

spinach

milk formula

liver

oily fish

oranges

kiwi fruit

FIGURE 1.2.6c: Which of these foods could treat which deficiency diseases?

Education and information ▶▶▶

The Department of Health provides information about the vitamins and minerals we need. Some people take vitamin and mineral supplement tablets rather than eating a balanced diet. This can be dangerous because some vitamins and minerals, such as vitamin A and iron, can cause health problems when taken in high doses.

6. Suggest why it is better to eat fresh fruit and vegetables as a source of vitamin C than to take tablets.

7. Design an engaging poster to educate primary school children about scurvy, rickets and anaemia. Refer to the symptoms and how each disease can be avoided.

Key vocabulary

deficiency disease

vitamin

scurvy

rickets

anaemia

Understanding the human digestive system

- Identify the organs of the human digestive system.
- Explain the role of digestion.
- Analyse links between digestion and the circulatory system.

The food we eat has chemical energy stored in it. To make use of this energy, we must digest the food. **Digestion** starts at the mouth and finishes at the anus. Many organs of the body are involved in digesting our food along the way.

FIGURE 1.2.7a: Digestion of your food starts in your mouth.

The human digestive system

Food is broken down and passes through each of the organs in the **digestive system**.

1. List the organs of the digestive system in the order that food passes through.

2. Suggest what would happen if there were a blockage in the large intestine.

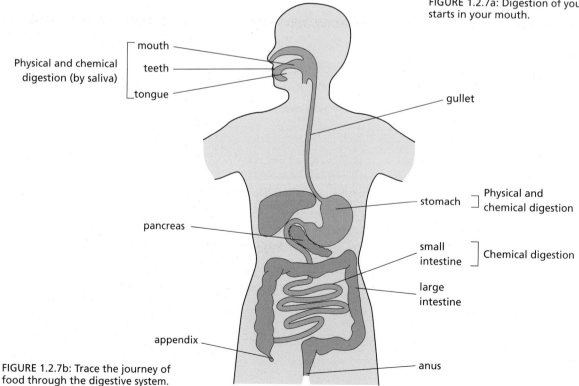

Physical and chemical digestion (by saliva)
- mouth
- teeth
- tongue

gullet

stomach — Physical and chemical digestion

pancreas

small intestine — Chemical digestion

large intestine

appendix

anus

FIGURE 1.2.7b: Trace the journey of food through the digestive system.

Why do we need digestion?

Food contains **chemical energy**. You must break down the large molecules in the food you eat into smaller molecules so that you can use the energy in your body. This process is called digestion.

The small food molecules are then absorbed and used in your body. Food that is not digested is passed out of your body. This waste is called faeces.

It takes up to a day for food to pass through a healthy digestive system. The length of time depends on the types of food that you have eaten.

3. Name the type of energy that food contains.

4. Explain why we need to break food into smaller molecules during digestion.

5. Describe what faeces are and explain what happens to faecal waste.

What happens to digested food next?

The small food molecules produced during digestion are absorbed into the bloodstream. This absorption takes place through the walls of the small intestine. The blood then carries the food molecules to the cells of the body where they can be used to release energy. This process of releasing energy from food molecules (and oxygen) is called **respiration**.

6. Describe how digested food molecules are carried around the body.

7. Complete the sentence by choosing the correct words in brackets.

    The walls of the small intestine must be (thin / thick) and have a (rich / poor) blood supply.

8. 'The digestive system and the **circulatory system** are linked.' Discuss this statement. Use the word respiration in your answer.

It takes quite a long time for your food to pass through your digestive system.

Time	Description
0 hours	Food is chewed and swallowed
1 hours	Food is churned with acids and enzymes in the stomach
2 hours	Partially digested food passes into the small intestine for further digestion and the start of absorption into the blood
6 hours	Undigested food passes into the large intestine and water is taken out of it and passes back into the blood
10 hours	The leftovers collect in the rectum
16–24hrs	The faeces pass out of the body

FIGURE 1.2.7c: How long is it since you last ate? Estimate where that food should be in your digestive system.

Did you know…?

Each day about 11.5 litres of digested foods, liquids and digestive juices flow through your digestive system.

Key vocabulary

digestion

digestive system

chemical energy

respiration

circulatory system

Investigating the start of digestion

We are learning how to:

- Describe what is meant by chemical and physical digestion.
- Explain how teeth and saliva are adapted to digestion.
- Suggest how results can demonstrate that digestion begins in the mouth.

Digestion begins as soon as food enters your mouth. Your teeth and saliva both have an important role and are adapted to do the job well. We can carry out some simple experiments to show how important the mouth is in digestion.

The role of the teeth ▶▶

As soon as food enters your mouth, your teeth start to crush it up. This mechanical breakdown of food is called **physical digestion**.

Physical digestion is important because:

- it makes food easier to swallow
- it makes it easier for chemicals to digest the food if it is already in smaller pieces.

We have different types of teeth that are *adapted* to break down food in certain ways.

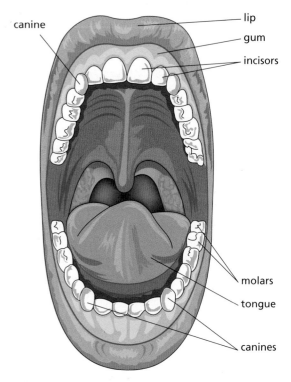

FIGURE 1.2.8a: How is each of the different types of teeth designed to break food down?

TABLE 1.2.8: Types of teeth and their uses

Type of teeth	Used for
incisors	biting and cutting
canines	ripping and tearing
pre-molars and molars	crushing and grinding

1. Describe what is meant by 'physical digestion'.

2. Give two reasons why mechanical breakdown is important in digestion.

3. Suggest which types of teeth are important to:
   a) grind salads
   b) tear meat

FIGURE 1.2.8b: Which teeth are especially important in eating this?

Why is saliva so important?

At the sight or smell of food, **saliva** production starts in your mouth. This is important because saliva helps to digest food in two ways:

- The liquid softens the food.
- It contains special chemicals that start to digest food.

The chemicals that break down large food molecules into smaller ones are called **enzymes**. An enzyme in saliva breaks down starch. So when you eat bread, pasta or rice, you start to break it down before the food is even swallowed. This type of digestion is called **chemical digestion**.

The teeth and saliva work together to produce a soft, round ball of food called a bolus. This wet bolus tastes good and is easy to swallow. Digestion can then continue throughout the rest of your digestive system.

4. Describe two functions of saliva in digestion.

5. Explain how saliva is adapted for digesting food.

6. Suggest two problems that you might have if you did not produce saliva.

The bread went soft.

It was wet as well.

The bread started to taste sweet.

FIGURE 1.2.8c: Have any of these students found evidence that digestion starts in the mouth?

How can we show that digestion starts in the mouth?

A group of students believed that digestion only starts in the stomach. Their teacher told them that the digestion of starch begins in the mouth. They decided to look for evidence. They chewed a piece of white bread for one minute before they swallowed it. Their observations are shown in Figure 1.2.8c.

7. Which of the students described some evidence that physical digestion had taken place?

8. Which of the students described some evidence that chemical digestion had taken place?

9. One student found out from the internet that starch molecules are broken down into glucose (sugar) molecules. Does this support the student with evidence that digestion starts in the mouth?

Did you know...?

We produce about 1.7 litres of saliva every day. That's more than the volume in four average cans of cola.

Key vocabulary

physical digestion

saliva

enzyme

chemical digestion

Understanding the roles of the digestive organs

We are learning how to:

- Describe the roles of the oesophagus, stomach, intestine and pancreas in digestion.
- Explain how the structure of each of the organs is adapted to its function.

Several organs in the body work together to digest food and dispose of waste. Food is churned, mixed, digested with chemicals and squeezed along the digestive system. Each organ has developed to carry out its specific role.

What do the organs of the digestive system do?

Digestion takes place along the digestive system from the mouth to the small intestine. Once digestion by enzymes is complete in the small intestine, the small molecule nutrients pass into the bloodstream and are carried to cells.

1. Name the organ in which digestion by enzymes is completed.

2. List two parts of the digestive system where physical *and* chemical digestion take place.

3. Food does not pass through the liver. Explain why it is shown as part of the digestive system.

Food enters the mouth where it is chewed up, rolled into a ball by the tongue and moistened by saliva ready for swallowing.

Food is swallowed and passes into the oesophagus which carries the food to the stomach.

The stomach breaks down food physically by muscle contraction and chemically by enzymes. The acid conditions kill bacteria and help the enzymes to work.

The small intestine digests the food further using different enzymes and absorbs it into the blood.

In the large intestine water is absorbed to make the waste (faeces) more solid.

The faeces are then passed out through the anus.

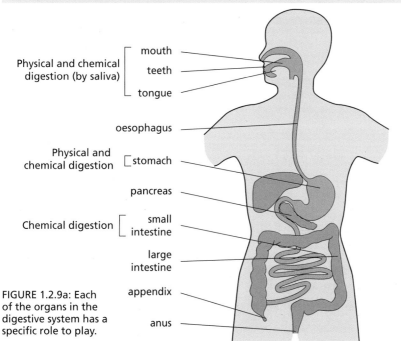

Physical and chemical digestion (by saliva) — mouth, teeth, tongue

oesophagus

Physical and chemical digestion — stomach

pancreas

Chemical digestion — small intestine

large intestine

appendix

anus

FIGURE 1.2.9a: Each of the organs in the digestive system has a specific role to play.

What about undigested food?

Not all of your food is digested. After digestion is complete in the small intestine, the undigested food travels through the large intestine where water is absorbed. The remaining waste is held in the rectum by a muscle, before we release it through the anus and into the toilet. This waste is called **faeces**. There are some foods we cannot digest, such as corn.

4. Where are faeces held before passing through the anus?

5. How could you tell that humans can't digest corn?

6. If the large intestine does not function properly, water is not completely absorbed. What condition could this lead to?

> **Did you know…?**
>
> In humans the small intestine is between 6 and 8 metres in length. That is more than the width of a typical swimming pool.

Adaptations of the organs

Each of the organs involved in digestion is specially designed for the job it does. This is known as an **adaptation**. For example, the mouth is well developed to chemically digest starch because it produces saliva.

TABLE 1.2.9: Adaptations of some of the digestive organs

Organ	How it is adapted to its function
oesophagus	Contains rings of muscle that contract behind the bolus to move the food along.
stomach	Contains muscles to squeeze the food. Secretes acid to kill bacteria. Contains an enzyme to digest protein.
pancreas	Releases enzymes that digest carbohydrates, protein and fats.
small intestine	Contains muscles to move the food along the tube and enzymes to complete digestion. Has thin walls and a good blood supply to help absorption of nutrients into blood.

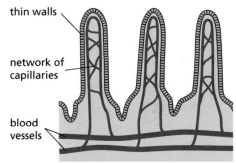

thin walls
network of capillaries
blood vessels

FIGURE 1.2.9b: Villi in the small intestine allow nutrients to be absorbed into the bloodstream.

7. How is the digestive system adapted to digest fats?

8. The stomach contains enzymes that work only in acidic conditions. Explain how the stomach is well adapted for this enzyme to work.

9. Look at Figure 1.2.9b. The villi show how the small intestine is folded to increase its surface area. Suggest how this increased surface area helps the role of absorbing nutrients into the blood.

Key vocabulary

faeces

adaptation

oesophagus

stomach

pancreas

small intestine

Applying key ideas

You have now met a number of important ideas in this chapter. This activity is an opportunity for you to apply them, just as scientists do. Read the text first, then have a go at the tasks. The first few are fairly easy, then they get a bit more challenging.

Can you stomach it?

Cows are sometimes said to have four stomachs. In fact, they have one stomach with four compartments. Each of the compartments has a different role and a different structure to support that role. Cows feed mainly on grass, which is actually quite difficult to digest because the fibre is not easily broken down. Grass does not provide much energy and so the digestive system of a cow must be as efficient as possible. Dairy farmers usually supplement the diet of their cattle with dried cereal, protein, vitamins and minerals.

FIGURE 1.2.10a: A cow needs four stomach compartments to digest grass.

The first stomach compartment contains bacteria that start to digest the fibre. The wall of the second compartment has a honeycomb structure. This compartment traps large food particles. These large particles are regurgitated, re-chewed and re-swallowed to digest them some more. The third compartment contains highly folded muscle to squeeze the water out of the food so that the water stays in the stomach. The fourth compartment is similar to the stomach of humans. It is acidic to allow enzymes to digest proteins.

FIGURE 1.2.10b: Snakes swallow their prey whole.

Snakes eat all parts of their prey. The function of their teeth is to capture the prey, rather than to grind food. They have powerful digestive enzymes to break down hair, feathers and bones.

Birds do not have teeth but use their beaks to tear food. Birds use a lot of energy to keep their bodies warm. So they must digest food as efficiently as possible. To help them to do this, they can hold food in their digestive system for a long time.

Some digestive systems are more efficient than others. The general relationship between the efficiency of digestion and the length of the digestive tract is the longer the tract, the more efficient digestion is.

Task 1: Sequencing the stages

Draw a flow chart to describe what happens in each of the four compartments of a cow's stomach. Highlight the compartment that is most like a human stomach.

Task 2: Cattle boosting

Think about the natural diet of a cow and explain why it needs such a complicated digestive system. Suggest why dairy farmers and beef farmers often give their cattle extra protein, vitamins and minerals.

Task 3: Explaining the adaptations

Discuss with a partner the role of each of the compartments of a cow's stomach. Discuss how each compartment is well designed for its role. Display your ideas in a table.

Task 4: Exploring other animals

Research some other animals with unusual digestive systems. Try to find how each system is well adapted to suit the needs of the animal.

Task 5: Looping it all together

Make a loop card game about digestion. You should write the questions on separate pieces of card. Write the answer to each question on different coloured pieces of card. Aim to write 10 questions and try to include ideas about digestion in several different species.

Task 6: Balancing speed and efficiency

Research has shown that animals that capture prey by a rapid chase often have a shorter digestive system than animals of similar size that do not rely on rapid acceleration to capture prey. Suggest which of the animals would have the most efficient digestion. Discuss the possible benefit to the animal that needs to accelerate rapidly in having a smaller digestive system.

Introducing enzymes

We are learning how to:

- Describe the role of different enzymes in digestion.
- Analyse a model of the digestive system.
- Explain observations of a practical activity to explore the role of enzymes.

Enzymes are protein molecules that occur naturally in your body. Without enzymes, we could not live because they speed up many vital processes, such as digestion.

Enzymes in digestion

If you eat a pizza, you need to digest the starch in the bread and the protein in the cheese. The enzyme **amylase**, released in the mouth and the small intestine, digests the **starch**. Starch is broken down into smaller molecules of **sugar** (glucose).

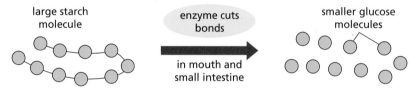

large starch molecule   enzyme cuts bonds   smaller glucose molecules

in mouth and small intestine

FIGURE 1.2.11b: Amylase enzyme breaks down starch to glucose molecules.

FIGURE 1.2.11a: You need enzymes to digest your pizza.

Protein is broken down by a different enzyme into smaller molecules, called amino acids. This enzyme is released in the stomach and the small intestine.

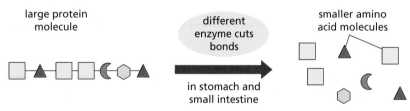

large protein molecule   different enzyme cuts bonds   smaller amino acid molecules

in stomach and small intestine

FIGURE 1.2.11c: Protein molecules are broken down to smaller molecules.

Fats are digested by another enzyme, lipase. This enzyme is released in the small intestine.

1. What are starch and protein broken down into during digestion?

2. List where each of the three enzymes that digest starch, protein and fats is found.

3. A pizza also contains fats. Fats are digested to fatty acids and glycerol by the enzyme lipase. Draw a string of triangles to represent a fat molecule, and show what happens to it when it is digested.

Enzymes are known as biological **catalysts** – they speed up reactions. If we didn't have enzymes, reactions such as respiration would go too slowly to keep us alive.

Enzymes are *specific*. This means that they can only break down one type of molecule. Enzymes are sometimes described as 'chemical scissors' because they cut up large food molecules into smaller molecules.

4. Describe what would happen if protein was mixed with amylase. Explain your answer.

5. Explain why we need enzymes to act as 'chemical scissors' during digestion.

6. Sufferers of cystic fibrosis release very few enzymes into the small intestine. What are the consequences of this?

> ### Did you know…?
>
> Enzymes are added to washing powders to digest stains from protein and fats on clothes. They are also used in the production of cheeses, yogurts and wines.

Modelling digestion »»»

A group of students investigated a model of digestion in the small intestine. Their set-up is shown in Figure 1.2.11d.

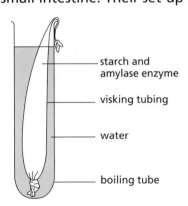

- starch and amylase enzyme
- visking tubing
- water
- boiling tube

FIGURE 1.2.11d: What does each part of the digestive system model represent?

The students mixed starch and amylase inside visking tubing. After 20 minutes, they tested the water outside the tubing for starch and glucose. They found that the water tested negative for starch but positive for glucose.

7. Describe what happened to the starch inside the tubing.

8. In the model, what was represented by the:

   a) visking tubing?

   b) starch inside the visking tubing?

   c) water surrounding the visking tubing?

9. Explain why glucose passed through the visking tubing but the starch did not.

Key vocabulary

amylase

starch

sugar

catalyst

Recognising the role of bacteria

We are learning how to:

- Describe the role of bacteria in our digestive system.
- Explain how the natural flora of bacteria can be disturbed.
- Analyse data about the effects of antibiotics on gut bacteria.

Scientists think that humans have over 500 different types of bacteria living in their digestive system. Most of the time, these bacteria do not harm us, indeed some help us to take the nutrients from the food that we eat during digestion.

How bacteria help with digestion

It is normal to have bacteria in your digestive system; the huge number and range found throughout your **gut** make up your gut **flora**. Many of these bacteria help with digestion.

Bacteria are found throughout the gut, but the number and type vary in different parts and is highest in the large intestine.

The helpful bacteria have several roles in digestion:

- They release energy from some foods such as sugars by fermentation.
- They make some enzymes that we need.
- They help to protect us from disease-causing bacteria.

Fermentation allows us to gain more energy from foods that have not been digested in the small intestine. The process also produces gases such as nitrogen, carbon dioxide and methane. The gas is passed out as 'wind'.

FIGURE 1.2.12a: Bacteria can be clearly seen in this magnified view of the inner surface of a healthy large intestine.

1. State where in the digestive system most bacteria are found.

2. Describe three ways in which some bacteria can help in digestion.

3. Suggest why few bacteria are found in the stomach.

> **Did you know…?**
>
> The digestive system is sterile at birth, but bacteria start to colonise the gut within a few hours of birth. By the time we are adults we should have up to 2 kg of bacteria in our gut.

It's all in the balance

Unfortunately, not of all of the bacteria in your gut are helpful – some may cause disease.

If you are healthy and eat fresh fruit and vegetables, the helpful bacteria prevent the disease-causing bacteria from harming you. The two types of bacteria can exist together and you come to no harm.

However, if the number of helpful bacteria is reduced, the disease-causing bacteria can suddenly increase in number. This can cause gastroenteritis, with symptoms such as diarrhoea, vomiting and nausea. Food such as **probiotic** yoghurt, which contains lots of helpful bacteria, will boost your helpful bacteria.

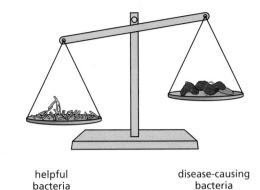

helpful bacteria   disease-causing bacteria

FIGURE 1.2.12b: In healthy people, the helpful bacteria outweigh the disease-causing bacteria.

4. Describe the balance between helpful and disease-causing bacteria in a healthy person.

5. How can we have disease-causing bacteria living in our gut without becoming ill?

6. Diarrhoea is associated with passing of lots of watery faeces. Suggest why this is a problem, particularly in young children.

What does the data show?

A group of scientists investigated the effect of **antibiotics** on some of the helpful bacteria.

TABLE 1.2.12: Effect of antibiotics on gut flora

Time after antibiotics taken (days)	0	0.25	0.5	1.0	2.0	4.0	6.0
Number of bacteria per gram of faeces ($\times 10^{11}$)	6.1	6.0	0.2	0.1	0.1	0.2	3.0

7. Describe the pattern in the results in Table 1.2.12.

8. Explain why the number of bacteria decreased during the investigation.

9. A doctor has recommended that a patient should drink some probiotic yoghurts, while taking antibiotics, to replace the helpful bacteria. When would be the most effective time for the patient to have the drinks?

Key vocabulary

gut

flora

probiotic

antibiotic

Understanding how we breathe

We are learning how to:

- Describe the mechanism of breathing in and out.
- Evaluate a model of breathing.
- Calculate changes in pressure and explain how these help us breathe.

A breathing system is important because it gets gases that we need into the body and moves waste gases out. Breathing is something that we do without even thinking about it. The brain controls movements in the chest which cause us to breathe in and out.

The mechanism of breathing

The main organ of the breathing system is the **lungs**. These are housed in the chest cavity.

Movements of your ribcage and **diaphragm** bring about breathing in and out.

1. Give another name for the windpipe.

2. Describe which way the diaphragm moves as you breathe in.

3. Describe the changes in the volume of the chest as you breathe in and then out.

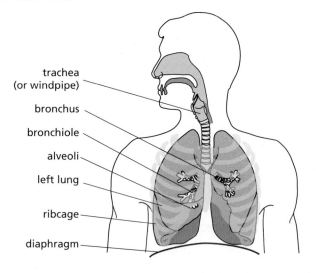

trachea (or windpipe)
bronchus
bronchiole
alveoli
left lung
ribcage
diaphragm

FIGURE 1.2.13a: Trace the journey of air through your nose to the alveoli.

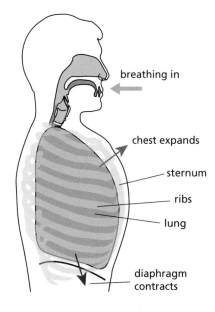

breathing in
chest expands
sternum
ribs
lung
diaphragm contracts

FIGURE 1.2.13b: Movements of the ribcage and diaphragm cause changes in the volume of the chest.

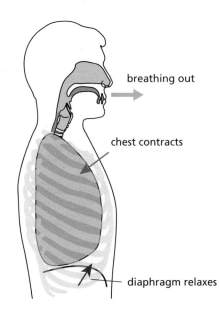

breathing out
chest contracts
diaphragm relaxes

A group of students have been shown a model (Figure 1.2.13c) to help them understand how we breathe in and out.

4. In the model shown in Figure 1.2.13c, what represents each of the following?

   a) the lungs

   b) the ribcage

   c) the diaphragm

   d) the trachea

5. Describe what happens to the 'lungs' as the 'diaphragm' in this model moves down.

6. Evaluate the model by listing ways in which the model matches a real breathing system and ways in which it doesn't.

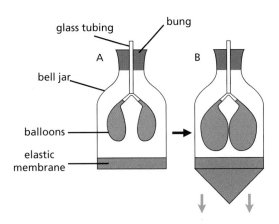

FIGURE 1.2.13c: You can make a model of the breathing system using a glass bell-jar and balloons.

Under pressure ⟩⟩⟩

A difference between the **pressure** inside the lungs and the pressure of the air around us (atmospheric pressure) is what causes us to breathe in and out.

- When you breathe in, the pressure in your lungs falls below atmospheric pressure and air automatically rushes into the lungs.

- When you breathe out, the pressure in your lungs rises above atmospheric pressure and air automatically rushes out of the lungs.

7. Describe when pressure in the lungs is at its highest and its lowest.

8. Suggest what would happen if the pressure in the lungs stayed the same as atmospheric pressure.

9. The pressure in the lungs just before breathing in has been calculated as −0.4 kilopascals (kPa). The pressure in the lungs just before breathing out has been calculated as 0.4 kPa. Calculate the pressure difference in the lungs between breathing in and breathing out. (Remember to use units.)

> **Did you know…?**
>
> Hiccups are caused by an involuntary contraction of the diaphragm. Hiccups are harmless and usually last only a few minutes. However, there are conditions under which hiccups persist for longer than a month.

Key vocabulary

lungs

diaphragm

pressure

Measuring breathing

We are learning how to:

- Describe what is meant by lung volume and identify some simple methods to measure it.
- Identify independent, dependent and control variables in a lung–volume investigation.
- Interpret and evaluate data linked to lung volume.

Lung volume can be measured quite easily and then be compared for different people. However, investigations involving people are often quite tricky to control and we must be aware of any potential errors.

Measuring lung volume 》》

The amount of air that you can breathe out following a big breath in is known as your **vital capacity** or your lung volume.

There are several ways of measuring lung volume, for example by displacing water.

The average lung volume of an adult male is 6 litres (L).

1. Describe what is meant by vital capacity.

2. A boy used the apparatus in Figure 1.2.14a to measure his lung volume. At the start, the bottle was full. Use the diagram to estimate the lung volume of the boy.

3. Suggest why the boy's lung volume is less than that of an adult male.

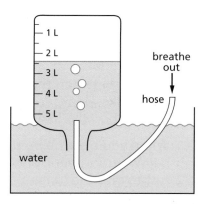

FIGURE 1.2.14a: As you breathe out strongly through the tube, water is displaced. The volume of water displaced is equal to the lung volume.

Investigating lung volume 》》》

A group of students wanted to find out whether being taller means that you have a bigger lung volume. They measured their own lung volumes by blowing into lung volume bags.

Their teacher gave them some definitions:

- **Independent variable** – the variable that we change in an investigation.

- **Dependent variable** – the variable that we measure in an investigation.

- **Control variables** – other factors that we need to control during the investigation.

4. Identify the dependent variable in the lung volume bag investigation. What units would it be measured in?

5. Identify the independent variable in the investigation. What units would it be measured in?

6. One student suggested that they should all take a deep breath in before measuring their lung volume. Suggest two other factors that the students would need to control.

Unravelling the data

Data was collected from some adult male students at a university to investigate the link between height and lung volume – the results are shown in Figure 1.2.14b.

7. Describe what this data shows about the link between height and lung volume.

8. From further investigation, the scientists found out that one of the men is an outstanding athlete. How tall is this man?

9. Suggest why this data may be more reliable than the data collected by the school students.

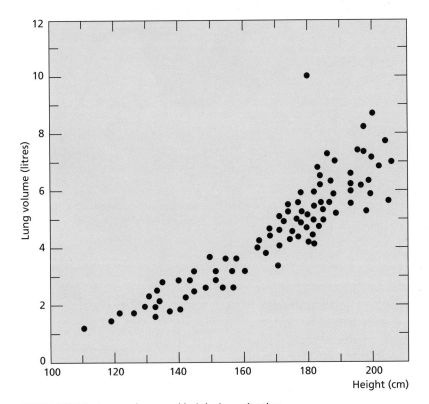

FIGURE 1.2.14b: Lung volume and height investigation.

Did you know…?

Spanish cyclist (and five times Tour de France winner) Miguel Indurain had lungs so big that they displaced his stomach, leading to a trademark paunch. His lung capacity was 8 litres when he was at his fittest.

FIGURE 1.2.14c: Miguel Indurain

Key vocabulary

vital capacity

independent variable

dependent variable

control variable

Evaluating gas exchange in humans

We are learning how to:

- Describe the features of the human gas exchange system.
- Explain how the features enable gases to be exchanged.
- Evaluate how well adapted the human gas exchange system is to its function.

Gases pass from the lungs to the blood and vice versa across the alveoli. These must have certain design features to enable this to happen efficiently. Without well-adapted alveoli we would be deprived of oxygen.

Our gas exchange system

Once in the lungs, air travels through the bronchioles to the **alveoli**. Gas exchange then takes place across the walls of the alveoli.

- Oxygen passes from the alveoli into the blood, which carries it to all the body cells where it is used in **respiration**.

- Carbon dioxide is a waste product of respiration in cells. It is carried by blood back to the lungs. Carbon dioxide passes from the blood into the alveoli. The carbon dioxide is breathed out.

1. Name the gas that passes back into the alveoli after respiration in the cells of the body.

2. Explain why oxygen is taken to all the cells in the body.

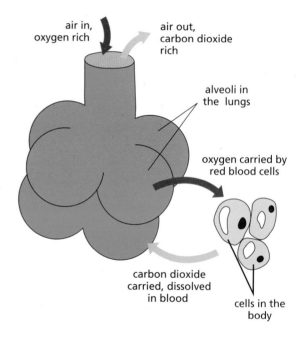

FIGURE 1.2.15a: Gas exchange across alveoli. Why do you think the alveoli have such a large surface area?

Labels: air in, oxygen rich; air out, carbon dioxide rich; alveoli in the lungs; oxygen carried by red blood cells; carbon dioxide carried, dissolved in blood; cells in the body

Designed to perfection?

Our gas exchange system has specific features that allow it to work efficiently. Tiny blood vessels called **capillaries** run over the surface of the alveoli. Gases travel between the blood in the capillaries and the air in the alveoli. This exchange is made as effective as possible by the thin lining of the alveoli and the very thin lining of the capillary walls. The surfaces of the alveoli are 'bumpy' (similar to cauliflower florets) and the walls are moist.

3. Describe what the surface of the alveoli is like.

4. Gases pass across moist surfaces more easily than across dry ones. Which gases does this help in the alveoli?

5. The thin surface of the alveoli means that the gases do not have to travel far between alveoli and blood. Explain how each of these other features of the alveoli supports gas exchange:

   a) moist surface

   b) surrounded by many blood capillaries

   c) large surface area.

Did you know…?

If the alveoli in one human's lungs were laid out, they would cover the area of a tennis court (100 m^2).

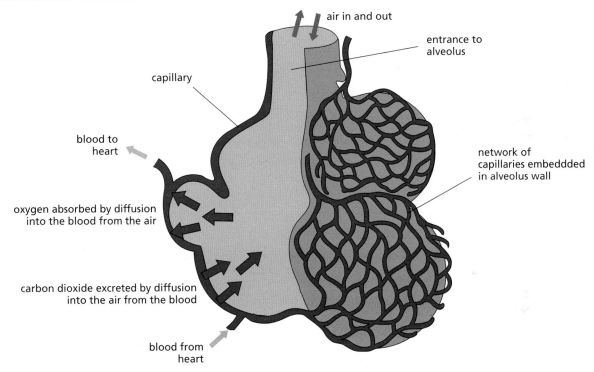

air in and out

entrance to alveolus

capillary

blood to heart

network of capillaries embeddded in alveolus wall

oxygen absorbed by diffusion into the blood from the air

carbon dioxide excreted by diffusion into the air from the blood

blood from heart

FIGURE 1.2.15b: The alveoli are surrounded by a huge network of capillaries.

Respiration or breathing? »»»

Breathing is the process of getting gases in and out of the body. Once the air is in the body, we use the oxygen in cells in respiration. Carbon dioxide gas is produced in cells by respiration. It travels to the lungs and is then breathed out.

6. Which parts of the body does breathing involve?

7. Which parts of the body does respiration involve?

8. A student has labelled a diagram of the breathing system (like Figure 1.2.13a) as the 'respiration system'. Explain why this is incorrect.

Key vocabulary

alveoli

respiration

capillary

Investigating diffusion

We are learning how to:

- Explain how diffusion makes breathing possible.
- Observe the effects of diffusion.
- Apply diffusion to our breathing system and ask questions to develop understanding.

Air particles are constantly moving because they have energy. As they move, they bump into each other and spread out. Diffusion allows us to smell food cooking, or a familiar person next to us! The gas exchange system also makes good use of diffusion.

FIGURE 1.2.16a: Why does the colour gradually spread evenly throughout the liquid?

What is diffusion?

As you discovered in Chapter 1, diffusion is the spreading out of **particles** from a region of high **concentration** to a region of low concentration. Sometimes this occurs across a **semi-permeable membrane**, which allows *some* particles to pass across it. If it is separating particles at different concentrations, the particles move across until they are evenly spread.

1. Describe what happens to particles during diffusion.

2. How does diffusion allow you to smell cooked food?

3. Suggest what happens when particles of equal concentration are on both sides of a semi-permeable membrane.

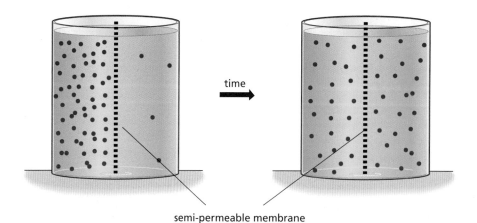

time

semi-permeable membrane

FIGURE 1.2.16b: Particles diffuse across a semi-permeable membrane.

Diffusion of gases occurs across alveoli. Oxygen from the air diffuses from the alveoli into the blood, and waste carbon dioxide diffuses from the blood into the alveoli.

In order for diffusion to occur efficiently, the walls of the alveoli and the blood capillaries must be thin and moist. Oxygen and carbon dioxide dissolve in the moisture of the alveoli and diffuse much more efficiently than across a dry surface.

4. What is the semi-permeable membrane in the lungs that gases diffuse across?

5. List the features of the surface of the alveoli that support efficient diffusion of oxygen.

6. Suggest what would happen if the surface of the alveoli was dry.

How does a concentration gradient help? >>>

Maintaining a difference in the concentration of particles is vital in diffusion. In the breathing system, the difference in concentration of oxygen between alveoli and blood is maintained by the blood constantly taking oxygen away. The difference in concentration between one side of a membrane and the other is known as the *concentration gradient*.

7. Where is the oxygen taken after it enters the blood across the alveoli?

8. Suggest the relationship between a concentration gradient and the rate of diffusion.

Key vocabulary

particle

concentration

semi-permeable membrane

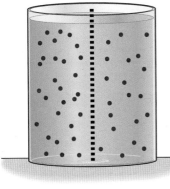 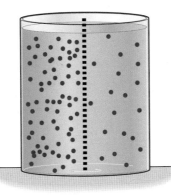 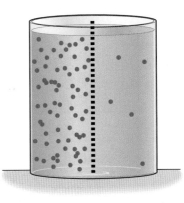

| low | medium | high |

FIGURE 1.2.16c: Compare the concentration gradients in the diagrams. This affects the rate of diffusion.

Exploring the effects of disease and lifestyle

We are learning how to:

- Describe the physical effects of disease and lifestyle on the breathing system.
- Explain the physical effects of disease and lifestyle on the breathing system.
- Describe how our understanding about the effects of smoking has changed over time.

The breathing system can be affected by both lifestyle choices and disease. We can't choose about inheriting a disease, but we do have a choice about whether or not we exercise or smoke.

Lifestyle choices and disease

Regular exercise improves breathing. It increases the strength of intercostal muscles so the chest can expand more when you breathe in, so your lung volume increases. It also increases the number and size of the blood capillaries surrounding the alveoli, so more oxygen can reach cells for respiration.

Diseases such as **asthma** affect the breathing system negatively. Asthma is thought to be influenced by inheritance, being born prematurely and being exposed to smoking in the womb or as a young child.

1. Give examples of a lifestyle choice and a disease that affect the breathing system.

2. Describe two ways in which the bronchioles change during an asthma attack, making it hard to breathe.

3. Describe two ways in which exercise affects the breathing system.

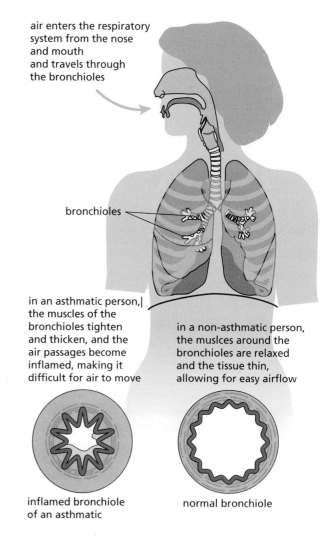

air enters the respiratory system from the nose and mouth and travels through the bronchioles

bronchioles

in an asthmatic person, the muscles of the bronchioles tighten and thicken, and the air passages become inflamed, making it difficult for air to move

in a non-asthmatic person, the muslces around the bronchioles are relaxed and the tissue thin, allowing for easy airflow

inflamed bronchiole of an asthmatic

normal bronchiole

FIGURE 1.2.17a: Asthma makes it hard to breathe.

Cigarettes release chemicals such as **nicotine**, **tar** and carbon monoxide.

- Nicotine is addictive.

- Carbon monoxide reduces the amount of oxygen the blood can carry.

- Tar can cause cancer.

In normal bronchioles, there are cells with tiny hairs called **cilia**. These sweep out any dirt particles that could damage the lungs.

In a smoker, the cilia are stuck together by tar and so dirt and smoke particles enter the lungs, which become irritated leading to a smoker's cough. Persistent coughing can damage the alveoli.

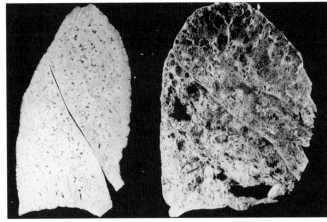

FIGURE 1.2.17b: Sticky brown tar coats the lungs of smokers. Tar can lead to lung cancer – the tobacco smoker's lung is on the right.

4. Name three dangerous chemicals in cigarettes.

5. Describe two effects of tar on the lungs.

6. How may smoking cause damage to people around the smoker?

Evidence takes time

We take it for granted now that smoking is bad for us, but it wasn't always like that. Before the 1950s, there were no health warnings with cigarettes and many believed they did no harm. In fact, cigarettes were glamourised by film stars.

In the 1950s, scientists recognised the link between smoking and lung cancer. However, it took some time before their theories were generally accepted. Cigarette adverts were later banned on TV, radio and in cinemas.

Only in the 1970s was it proven and accepted that smoking can harm the people around a smoker as well as the smoker, and affects a developing foetus during pregnancy.

7. Suggest the role that research took in cigarette advertising being banned.

8. Suggest why some people do not stop smoking even though it damages health.

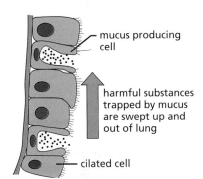

mucus producing cell

harmful substances trapped by mucus are swept up and out of lung

cilated cell

FIGURE 1.2.17c: What would happen without cilia?

Did you know…?

Research estimates that, on average, each cigarette shortens a smoker's life by 11 minutes.

Key vocabulary

asthma

nicotine

tar

cilia

Checking your progress

To make good progress in understanding science you need to focus on these ideas and skills.

- Describe the components of a healthy diet (food groups).
- Recall the tests for starch and sugar.
- Suggest some foods that contain starch and sugar.

- Explain the role of some of the components of a healthy diet.
- Recall the tests for starch, sugar, protein and fats.
- Suggest several foods that contain starch, sugar, proteins and fats.

- Explain the role of all of the components of a healthy diet.
- Predict the observations of food tests on several foods for starch, sugar, protein and fats.

- List groups of people who need different amounts of energy from food.
- Describe some of the physical effects of obesity and starvation.
- Describe the cause and symptoms of scurvy and suggest foods to treat it.

- Compare the energy requirements of different people – such as men and women, teenagers and the elderly, pregnant and non-pregnant women.
- Explain some of the physical effects of obesity and starvation.
- Describe the causes of several deficiency diseases and suggest foods to treat each.

- Explain why different groups of people have different energy requirements.
- Use data on packaging to plan how individuals could meet their energy requirements.

- Name some of the organs of the digestive system.
- Describe what is meant by physical digestion and chemical digestion.

- Locate the organs of the digestive system on a diagram.
- Recall where physical digestion takes place and where chemical digestion takes place.
- Explain how teeth and saliva are adapted to digest food.

- Name the organs of the digestive system in the order that food passes through them.
- Compare and contrast physical and chemical digestion.
- Explain the link between digestion and circulation.

- Describe the role of the stomach and small intestine in digestion.
- Recall the names of some digestive enzymes.

- Describe the role of all the organs in the digestive system.
- Describe some adaptations of the organs in the digestive system.
- Explain the role of three digestive enzymes.

- Explain how the structure of each of the organs in the digestive system supports its function.
- Explain how visking tubing can be used to model the digestive system.
- Apply the role of each of the digestive enzymes to different foods.

- Describe the movement of the ribs and diaphragm during breathing in and out.
- Describe what is meant by lung volume.

- Explain how changes in pressure in the chest bring about breathing in and out.
- Describe two ways of measuring lung volume.

- Compare the pressure in the chest before breathing in and breathing out with atmospheric pressure.
- Compare two ways of measuring lung volume.

- Recall which gas from the air is used in the body.
- Describe where gases are exchanged between the lungs and the blood.
- Describe examples of disease and lifestyle choices that affect the breathing system.

- Describe four features of the alveoli that help gas exchange.
- Explain how each feature of the alveoli supports gas exchange.
- Describe the effects of exercise, asthma and smoking on the breathing system.

- Apply the structure of alveoli to their function in gas exchange.
- Explain the effects of exercise, asthma and smoking on the breathing system.
- Explain the difference between breathing and respiration.

Questions

See how well you have understood the ideas in the chapter.

1. What is the main food group found in fish? [1]

   a) carbohydrate   b) fibre   c) protein   d) fat

2. Which food group adds bulk to our food and helps to move it through the digestive system? [1]

   a) fibre   b) fats   c) vitamins   d) protein

3. Which organ in the digestive system contains hydrochloric acid? [1]

   a) mouth   b) small intestine   c) large intestine   d) stomach

4. Which gas is there more of in exhaled (breathed out) air than inhaled (breathed in) air? [1]

   a) oxygen   b) carbon dioxide   c) nitrogen   d) sulfur dioxide

5. Describe *two* ways in which we use energy when we are sleeping. [2]

6. Describe the changes in the volume and pressure inside the chest just before breathing in. [2]

7. Explain, using examples, the differences between chemical digestion and physical digestion. [4]

See how well you can apply the ideas in this chapter to new situations.

8. Which would be a good reason to check food labelling? [1]

   a) Starch turns iodine blue/black.

   b) It could help to avoid an allergic reaction.

   c) Proteins are broken down by enzymes.

   d) Energy is measured in kilojoules.

9. Why is it difficult to estimate the number of people dying from obesity? [1]

   a) The number is increasing.

   b) The number is decreasing.

   c) People with obesity often die from other conditions.

   d) Obesity doesn't kill people.

10. Which of the following foods would the pancreas help most in digestion? [1]

    a) turkey breast   **b)** lettuce   **c)** apple   **d)** butter

11. Figure 1.2.19a shows particles in a container with a semi-permeable membrane. How would the diagram look if it showed the same container after 24 hours? [1]

    a) The diagram would not change.

    b) The diagram would show all particles on other side of membrane.

    c) The diagram would show the same number of particles evenly spread on both sides.

    d) The diagram would show many more particles evenly spread on both sides.

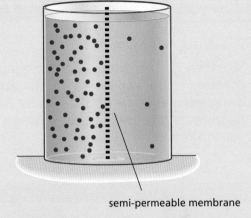

semi-permeable membrane

Figure 1.2.19a

12. Suggest why smoking is even more harmful to an asthma sufferer than to someone without asthma. [2]

13. Explain why lemons can help to treat a person with scurvy. [2]

14. Both the alveoli in the lungs and the villi in the small intestine are well adapted to allow materials to pass through. Compare the role of each and how it is adapted to its role. [4]

Questions 15–16

See how well you can understand and explain new ideas and evidence.

15. Probiotic drinks are very popular as a way to increase the numbers of certain bacteria in our gut. Suggest why these drinks are important after taking antibiotics. [2]

FIGURE 1.2.19b

16. A diet company claims that it has produced a medicine that blocks lipase (fat-digesting) enzymes. Explain how such a medicine may support weight loss. [4]

Mixing, Dissolving and Separating

Ideas you have met before ⟩⟩⟩

Carrying out experiments

Scientists gather evidence by carrying out experiments. For the evidence to be useful, the experiment has to be designed and carried out carefully. Results and observations need to be recorded so that patterns can be seen and conclusions drawn.

Heating a liquid

If we heat water enough it will boil and turn into steam. Steam is a gas that, if it cools down, will turn back into a liquid.

Drying involves evaporation, which occurs faster with heat and the movement of air.

Dissolving

Some materials – such as salt and sugar – can dissolve in water. We say that these are soluble.

Other materials – such as sand – do not dissolve in water. We say that these are insoluble.

Separating mixtures

If something has been mixed with water but has not dissolved, we can separate it by using a filter or a sieve.

This method can be used to remove sand and gravel from water, but filtering does not remove soluble substances such as salt.

In this chapter you will find out

Using laboratory equipment

- A Bunsen burner can be used to supply heat to speed up a chemical reaction or to cause a change of state.

- Measuring the mass before and after a reaction helps us to understand some important ideas in chemistry.

Distillation

- If we heat a liquid it will evaporate, turning into a vapour (gas). If we then cool the vapour, it will turn back into a liquid.

- If we heat different liquids, we find that they boil at different temperatures.

- We can use this information about boiling points to separate mixtures of liquids. The process is called distillation and is used to make perfume and also fuels such as petrol.

Solubility

- When substances dissolve we can explain the process by thinking about the particles they are made of.

- Dissolving often happens more quickly at higher temperatures because of the extra energy of the particles.

- Solutions are very useful.

Chromatography

- Soluble substances can be made to travel up filter paper by adding a solvent.

- If we do this with coloured dyes or inks, we find that the different colours in the mixture move different distances.

- This technique is called paper chromatography and can be used to separate mixtures and identify chemicals.

Working safely in a laboratory

We are learning how to:

- Recognise and reduce risks when working in the laboratory.
- Name and select appropriate equipment.

Many activities that we do are risky – such as crossing the road or playing sport. That does not mean we do not do them, but it is important that we take precautions to reduce the likelihood of injury.

Staying safe

In many sports – such as cycling, cricket and horse riding – athletes are encouraged to wear head protection. Some jobs require employees to wear special clothing, head, ear or eye protection to minimise the **risk** of injury and keep them safe. Safety is very important in a science **laboratory**.

1. State three jobs that require workers to wear hard hats.

2. Why do doctors and nurses often wear gloves?

3. **a)** What safety equipment is the rollerblader wearing?

   b) List other safety equipment that a competing rollerblader might wear, and how it would protect him/her.

FIGURE 1.3.2a: Head protection makes rollerblading safer, but no less fun.

Safety in the laboratory

Ignoring **hazards** can lead to accidents and people being injured. If we identify and reduce the risk of these hazards then we can work safely. Hazards in the laboratory can come from chemicals, glass equipment or hot objects. The way we behave can also affect the risks to ourselves and to others. Wearing safety goggles is important when performing experiments to protect your eyes from splashes and objects that may splinter or produce sparks.

FIGURE 1.3.2b: Students in this laboratory are NOT safe.

4. How many hazards can you identify in Figure 1.3.2b?

5. How could the risk from these hazards be reduced?

6. Write your top ten safety rules for working safely in the laboratory.

Laboratory apparatus

The laboratory, like a kitchen, contains equipment for heating, measuring, mixing and pouring. However, the names for the equipment are different. It is important to use the correct scientific names for practical **apparatus** so that other scientists can copy your methods to compare and verify the results.

Scientists use a **Bunsen burner** to heat substances to very high temperatures. Knowing how to use the burner safely is very important to reduce the risks of injury and fire.

A Bunsen has two colours of flame depending on how much air is mixed with the methane gas before it is burned, as shown in Figure 1.3.2c. This is controlled by moving the collar to open and close the air hole at the base. We always leave Bunsen burners with the air hole closed when they are not being used.

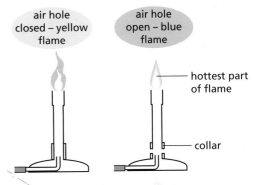

FIGURE 1.3.2c: The flame on a Bunsen burner can be controlled.

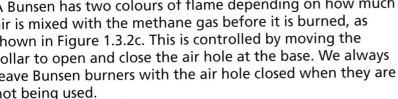

7. If you were using a Bunsen burner to heat a liquid in a beaker, why would it be best to:

   a) use a glass beaker, rather than a plastic one?

   b) use a tripod and gauze?

8. Explain how the following apparatus reduces risks in laboratory experiments:

   a) test-tube rack

   b) clamp and stand.

9. Explain how you could reduce the risks when heating a substance in a boiling tube.

10. Give as many advantages as you can for using a heatproof bench mat while heating substances directly in a flame using tongs.

Did you know...?

25 000 children under 5 years old attend hospital in the UK every year after being accidentally poisoned by substances in their own home.
(*Royal Society for the Prevention of Accidents*)

Key vocabulary

risk

laboratory

hazard

apparatus

Bunsen burner

Recording experiments

We are learning how to:

- Represent scientific experiments clearly.
- Make and record accurate measurements.

Symbols and simple diagrams can accurately represent an object, message or procedure, without requiring the skills of an artist.

Representing apparatus

It is important to be able to represent how an **experiment** has been carried out clearly so that others can understand what has been done. This can be achieved using a simple, common series of 2D images.

An evaporating basin, for example, is represented with a simple 2D **line diagram** (see Figure 1.3.3b). Notice that there is no top line. This shows that the dish is open and has no lid or cover.

You can use the same style of diagram to show how to set up all the apparatus for an experiment.

1. Why are simple 2D diagrams used to represent equipment?

2. Why do scientists use an arrow to represent a Bunsen burner?

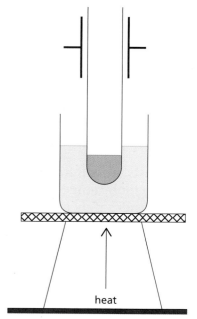

FIGURE 1.3.3a: This type of 2D diagram can be drawn to illustrate apparatus.

FIGURE 1.3.3b: An evaporating basin and a simple 2D diagram representing it

Measuring accurately

Measuring cylinders come in different sizes – for example 10 cm³, 25 cm³ – so that you can choose the one that most closely matches the volume you are **measuring**. This makes measurements more *accurate* by helping you measure closer to the volume needed.

FIGURE 1.3.3c: Measuring cylinders of different sizes

Liquids are not as easy to measure as they may seem because they do not form a straight line at the surface. The surface curves downwards – this is called a **meniscus**. Scientists use the bottom of the meniscus to make their readings. They do this at eye level so that they can be sure that they are reading the position and scale accurately.

Balances can measure mass to different numbers of decimal places. The more decimal places there are the more precise the measurement is and the more *sensitive* the balance is to very small differences in mass.

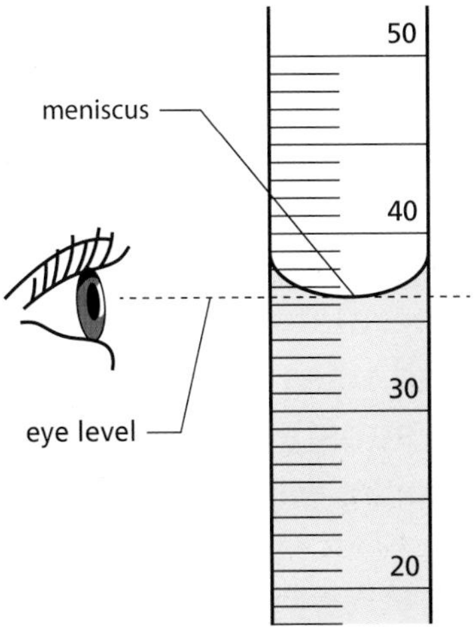

FIGURE 1.3.3d: Reading the meniscus at eye level

3. Explain your choice of equipment to make the following measurements:

   a) 6 cm³   **b)** 23 cm³   **c)** 10 g

   d) 42.40 g   **e)** 20 °C   **f)** 62 s

4. Why is it important to choose the appropriate measuring equipment?

Reporting results

Results are recorded in tables with headings, units and **data**. If the mass was measured on a balance with one decimal place, all the data should be reported with one decimal place, including the average result.

Mass of powder	Volume of gas (cm³)		
	First experiment	Repeated experiment	Average
1 g	5	7	6
10 g	90	95	92.5
5 g	27	23	25

FIGURE 1.3.3e: Example of a results table

5. Say why you think the table in Figure 1.3.3e is a good one and how it could be improved.

6. Write a checklist that your class could use to make sure that your results are recorded accurately.

7. The same 100 cm³ measuring cylinder was used to make all the measurements in the table. How might this have affected the results?

Did you know...?

The meniscus in mercury curves upwards. So, when you read the temperature on a thermometer, you should read the top of the meniscus at eye level.

Key vocabulary

experiment

line diagram

measuring

meniscus

data

Recognising materials, substances and elements

We are learning how to:

- Recognise the difference between materials, substances and elements.
- Identify elements by their names and symbols.
- Explain what is meant by a chemically pure substance.

Chemistry is the study of the structure, properties and uses of materials. Materials are combinations of substances that are made of basic building blocks – the chemical elements.

Elements and compounds

For a scientist, a **material** is anything made of matter or particles. Most materials are made up of combinations of chemical substances, called **compounds**.

Compounds are made up of different **elements**. Elements are the chemical building blocks of materials – each is made up of only one type of atom. Each element is identified by a unique **symbol**, which always begins with a capital letter. Table 1.3.4a lists some common elements and their symbols.

Some elements have symbols based on their old names, such as iron (ferrum), gold (aurum) and copper (cuprum).

1. What is the chemical symbol for carbon?

2. Explain why CO and Co are not the same thing.

Pure or not?

When elements combine in a compound, sometimes they form a **molecule**. A water molecule is made up of two hydrogen (H) atoms and one oxygen (O) atom. We write this as H_2O.

Chemically **pure** water contains only H_2O molecules. Bottled water contains other substances – for example, magnesium (Mg) and calcium (Ca). It is not the same as pure water.

3. Which elements make up the sugar called glucose ($C_6H_{12}O_6$)?

4. Describe the difference between chemically pure water and so-called 'naturally pure' bottled water.

FIGURE 1.3.4a: Orange juice may contain over 10 different compounds, including water, sugar and vitamins.

TABLE 1.3.4a

Name of element	Symbol
hydrogen	H
oxygen	O
carbon	C
cobalt	Co
iron	Fe
gold	Au
copper	Cu
calcium	Ca

5. Explain why calcium carbonate, $CaCO_3$, is a pure substance but not an element.

6. Why do you think that we call substances like orange juice and bottled water *pure*?

Elements in the body 〉〉〉

Almost 99 per cent of the mass of your body is made up of only six elements, but these are combined in different ways. Blood contains water, salts and an iron-containing compound called haemoglobin. Bones and teeth are made from compounds that contain calcium and phosphorus.

In total, your body contains around 28 different elements (Table 1.3.4b) whose atoms combine in various ways to create the different substances that you are made of.

TABLE 1.3.4b

Elements in the human body	Make up (%)
oxygen	65
carbon	18
hydrogen	10
nitrogen	3
calcium	1.5
phosphorus	1.0
potassium	0.35
sulfur	0.25
sodium	0.15
magnesium	0.05
copper, zinc, selenium, molybdenum, fluorine, chlorine, iodine, manganese, cobalt, iron	0.70
lithium, strontium, aluminium, silicon, lead, vanadium, arsenic, bromine	trace amounts

7. Explain why blood is not a pure substance but haemoglobin is.

8. Draw a graph to represent the 10 most abundant elements in your body. Include the symbols.

Did you know...?

Before recorded history, many elements were in use but only in the form of their compounds. It was not until the late 1700s that pure elements started to be isolated and identified.

Key vocabulary

material

compound

element

symbol

molecule

pure

Understanding water

We are learning how to:

- Recognise the importance and different sources of water.
- Explain the differences between types of water.

Approximately 70 per cent of the Earth is covered by water. It exists in oceans, rivers, glaciers and falls from the sky as rain and snow. We rely on it to drink and wash, but also to grow food and to transport us from place to place. It literally sustains life on our planet.

FIGURE 1.3.5a: Much of the Earth is covered by water.

Would you drink seawater?

Seawater tastes salty because of the minerals and salts dissolved in it. Seawater is a **mixture**. It contains substances dissolved from rocks and the atmosphere. Some elements and compounds can be extracted from it and are used to make other things including chlorine, bromine and iodine.

1. Why is seawater described as a mixture?
2. Suggest an easy way of proving that there are substances dissolved in seawater.

What is in our water?

Because **water** is good at dissolving substances, natural sources of water on Earth are mixtures. Even bottled mineral water is not pure water because it contains substances that have been dissolved from the rocks surrounding it. These are not harmful and, in fact, can be good for us.

Some tap water contains a lot of calcium compounds that sometimes appear as a white substance in kettles when the water is boiled – this is called **limescale**. Some of our tap water has fluoride added because it is good for our teeth. Other substances, such as chlorine compounds, are added to kill bacteria and make the water safe to drink.

	(mg/l)
Calcium (Ca)	181.0
Chloride (Cl$^-$)	57.5
Bicarbonate (HCO$_3^-$)	239.0
Fluoride (F$^-$)	0.5
Lithium (Li)	0.2
Magnesium (Mg)	53.5
Nitrate (NO$_3^-$)	2.2
Potassium (K)	2.5
Silica (SiO$_2$)	7.5
Sodium (Na)	36.1
Strontium (Sr)	3.2
Sulfate (SO$_4^{2-}$)	459.0

FIGURE 1.3.5b: Bottled water includes substances such as calcium. So does tap water.

3. Name two chemicals that might be added to water when it is being prepared to be supplied to us.

4. If there is calcium in the water we drink, how might it have got there?

5. Suggest why drinking water supplied in different parts of the country might taste slightly different.

The need for clean water »»»

Not all countries around the world have access to clean, safe water. In some parts of Africa, for example, many people die due to lack of water or from diseases caught from drinking dirty water. Some African countries are hot, barren and have had significant periods of drought where crops fail and food is scarce. By drilling wells deep underground and providing methods for cleaning and **purifying** water, the lives of people in these areas can be transformed.

FIGURE 1.3.5c: People need clean water.

Our water is recycled. We collect and use rainwater in reservoirs – even the water we flush or wash away is collected and cleaned, and then returned to the main water supply. In order to recycle water effectively, an understanding of chemistry is vital. If too much or the wrong chemical is used, many people would be affected and could be poisoned or become ill from bacteria in the water.

6. Why is a supply of clean water so critical?

7. What is the difference between clean water and pure water?

8. Suggest why the supply of clean water is improved:

    a) in parts of Africa by drilling deep wells

    b) in many countries by building reservoirs.

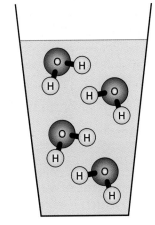

Did you know...?

Water, H_2O, should be a gas at room temperature, like hydrogen sulfide, H_2S. However, there are special forces between water molecules that hold them together more strongly than molecules in similar compounds.

FIGURE 1.3.5d: Water is a special case.

Key vocabulary

mixture

water

limescale

purifying

Dissolving

We are learning how to:

- Explain the terms solvent, solution, solute and soluble.
- Identify factors that affect dissolving.
- Explain the difference between a dilute solution and a concentrated solution.

Limestone caves are amazing places. Stalactites grow down from the roof and, where the water drips down and hits the cave floor, stalagmites grow upwards. They grow only a few centimetres every hundred years as water slowly deposits the minerals it dissolved when passing through the limestone rock.

FIGURE 1.3.6a: Even rocks dissolve.

Do you take sugar? 》》

If you stir sugar into a cup of tea or coffee the crystals disappear – they dissolve. The water is called the **solvent** and the mixture is called a **solution**. The sweeter the taste, the more sugar has dissolved. Substances that dissolve are described as **soluble**.

1. Why does sugar disappear when you stir it in tea?
2. How do you know that the sugar is still there in the drink?
3. What is a solution?

FIGURE 1.3.6b: The sugar seems to disappear but the tea tastes sweet.

Different sugars 》》》

Different kinds of sugar all contain sucrose, which is extracted from plants like sugar cane or sugar beet. Although the size of the crystals varies, the sucrose molecules ($C_{12}H_{22}O_{11}$) are all the same.

If you use a beaker containing water you can see how easily sugar dissolves. A solid that dissolves is called a **solute**. Not all types of sugar dissolve in water at the same speed.

FIGURE 1.3.6c: Does the type of sugar affect how it dissolves?

TABLE 1.3.6: The rate at which different types of sugar dissolve

Sugar type	Mass used (g)	Volume of water (cm³)	No. of stirs to dissolve it all
caster	1.50	100	20
white cubes	1.50	100	75
granulated	1.45	100	42
brown sugar	1.50	100	58

Look at the experiment results in Table 1.3.6 to answer these questions.

4. What were the solute and the solvent?

5. Put the sugar types in order of size of their sugar pieces, starting with the largest. What do you notice about how they dissolved?

6. Explain why the experiment was a fair test.

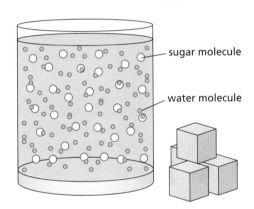

Did you know...?

The artificial sweetener aspartame is 250 times sweeter than sucrose. However, a natural sweet protein called thaumatin, found in the West African katemfe fruit, is 3000 times sweeter than sucrose!

Dissolving sugar

Sugar lumps are made of sugar crystals packed together. Each of these crystals contains many tiny sugar molecules. The solution in the glass in Figure 1.3.6d contains a mixture of water molecules (solvent) and sugar molecules (solute).

The water molecules are much smaller than the original sugar crystals and are able to break down the crystals into sugar molecules. The movement of the water molecules helps to separate and spread the sugar molecules throughout the solution.

The separated molecules are so small that they seem to have disappeared into the water. The mass of sugar dissolved in a particular volume of water is called the **concentration**. If there is a lot of sugar in the water it is a concentrated solution; if there is only a little it is a dilute solution.

sugar molecule

water molecule

FIGURE 1.3.6d: What happens when sugar dissolves?

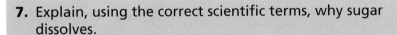

7. Explain, using the correct scientific terms, why sugar dissolves.

8. Why does stirring make the sugar dissolve faster?

9. Will icing sugar dissolve faster than caster sugar? Explain your answer.

10. Draw a diagram like the one in Figure 1.3.6d to show the difference between a concentrated solution and a dilute solution.

Key vocabulary

solvent

solution

soluble

solute

concentration

Separating mixtures

We are learning how to:

- Recognise the differences between substances and use these to separate them.

If you put different objects together, such as different fruits in a bowl, toys in a box or sweets in a bag, you have a mixture. Simply picking out one object separates it from the mixture. You can have mixtures of elements and compounds too. These are not so easy to separate.

Using size to separate mixtures

Gravel and rocks can be removed from sand by sieving. This separation depends on the size of the holes in the sieve. However, if the sand was mixed with water, this method would not work. A **filter** would be needed instead.

Filters are often made of paper or cloth with very small holes that are difficult to see without a microscope. Filters are often used to remove the solids when making coffee. Tea bags act as filters, whereas a tea strainer acts as a sieve. Air and fuel filters are used in cars to remove particles that would damage the engine.

FIGURE 1.3.7a: Using a sieve

1. What is the difference between a filter and a sieve?
2. Explain how filters and sieves are helpful when making tea and coffee.

Being different

Mixtures can be separated by finding differences between the substances. For example, there are only three metals that are attracted by a magnet – iron, cobalt and nickel. We can use this difference to separate these magnetic metals from mixtures.

FIGURE 1.3.7b: The physical property of magnetism can be used to separate magnetic from non-magnetic materials.

TABLE 1.3.7

Rules for mixtures	
1	Mixtures can be separated by physical methods.
2	Mixtures only have the properties of the things in the mixture.
3	Mixtures of elements can be made using different amounts of each one.
4	No chemical change occurs when making mixtures.

3. Would all of a mixture containing iron filings and lead powder be magnetic?

4. If nickel chloride were mixed with lead, could you use a magnet to separate them? Explain your answer.

5. Use the rules in Table 1.3.7 to explain why mixtures can be separated using known differences between the substances.

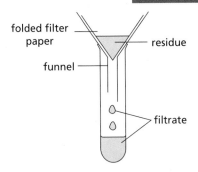

FIGURE 1.3.7c: Separating mixtures by filtration

Separation by filtration >>>

In the laboratory, filter paper can be used to separate solids from liquids – this process is called **filtration**. Substances that do not dissolve are described as **insoluble**. Filtration helps to separate soluble and insoluble solid substances.

Liquids like oil and water do not mix. The oil does not dissolve in the water to make a solution. These liquids are described as **immiscible**. The lighter oil floats on top of the water, and even if you shake the mixture, the two layers will reappear as the two liquids separate again.

The way these two liquids behave means that a separating funnel can be used to split them up. The water layer can be removed using the tap at the bottom, leaving the oil layer behind.

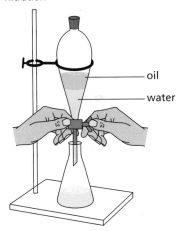

FIGURE 1.3.7d: Using a separating funnel to separate immiscible liquids

Did you know...?

The components of blood can be separated by spinning the blood really fast in a centrifuge. This causes the red blood cells to separate from the plasma because they are denser and sink to the bottom.

6. Choose a method to separate flour and rice.

7. Explain why a bottle of salad dressing made from vinegar and olive oil must be shaken before use.

8. Explain why filtration would not separate sugar and water.

9. Create a key or flow diagram to help explain which method of separation to use for a mixture.

Key vocabulary

filter

mixture

filtration

insoluble

immiscible

Dissolving and evaporating

We are learning how to:

- Separate a soluble substance from water.
- Form crystals from solutions.
- Explain solubility.

Gemstones, such as amethyst, are crystal that formed deep in the Earth's crust. Water dissolved salts and as this solution cooled over thousands of years, the precious crystals formed.

FIGURE 1.3.8a: Amethyst crystals formed naturally by the processes of dissolving and evaporating.

Temperature effects

One way to help things dissolve is to increase the temperature of the water. This is why we wash clothes in warm water. Any **soluble** stains in the clothes will dissolve better at a higher temperature. The mass of solute that dissolves in a solvent at a particular temperature is called its **solubility**.

Look at the data in Table 1.3.8. The results show the mass of sugar (sucrose) that can dissolve in 100 cm³ of water.

TABLE 1.3.8: Dissolving sugar in water at different temperatures

Temperature of water (°C)	0	20	40	80
Mass of sucrose that can dissolve (g)	180	200	240	600

1. What does the data in Table 1.3.8 tell you about the solubility of sucrose at different temperatures?

2. How could you display this data to show the pattern more clearly?

3. Estimate the mass of sucrose that will dissolve at 60 °C.

Did you know…?

These amazing natural gypsum crystals were found 300 metres underground in a mine in Mexico in 2000. They have grown undisturbed for thousands of years. Some are as long as 12 metres.

FIGURE 1.3.8b Naica gypsum crystals

Making crystals >>>

Heat can also be used to separate soluble substances from their solutions. When the solvent evaporates it leaves behind the solid solute – this is called **crystallisation**. If this process happens quickly, small **crystals** of the solute will form. However, if the evaporation happens slowly, bigger crystals can grow.

Crystallisation happens most efficiently when the solution is **saturated**. This means that there is as much solute dissolved in a solvent as possible. If any more solute is added to a saturated solution it will not dissolve. As the solvent cools, the crystals start to form. The solubility of substances depends on the temperature of the solvent.

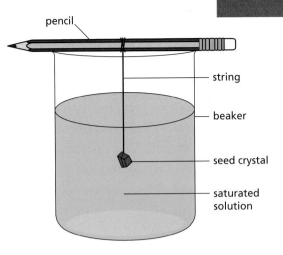

FIGURE 1.3.8c: You can grow your own crystals.

4. What is a 'saturated' solution?

5. Describe how you could grow salt crystals.

6. Why do you think that the crystals start to form as the solvent cools?

Using graphs >>>

Substances dissolve more in hot water because the water molecules have more energy and move faster. They can break down the crystals and separate the solute molecules more quickly. Solubility also depends on the type of solute. The graph shows the change in solubility of different salts with temperature.

7. Look at Figure 1.3.8d. Which salt is most soluble at 60 °C?

8. If 50 g of potassium nitrate were added to water at 20 °C, would it all dissolve? How do you know?

9. What would you see if a solution containing 50 g of sodium chloride was cooled from 100 °C to 40 °C?

10. How much sodium nitrate would you need to add to 100 g of water at 50 °C to make a saturated solution?

11. Using your knowledge of dissolving, explain why there is a connection between the temperature of a solvent and the solubility of a salt.

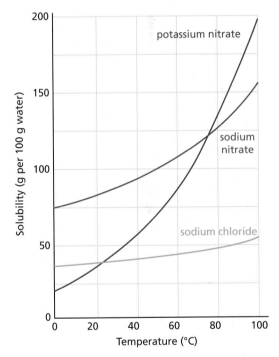

FIGURE 1.3.8d: Solubility graph

Key vocabulary

soluble

solubility

crystallisation

crystal

saturated

Extracting salt

We are learning how to:

- Identify sources of salt and describe how it is extracted.
- Recognise the uses and importance of salt.
- Obtain pure salt from a mixture.

Salt has always been valuable to flavour and preserve food. The chemical name for salt is sodium chloride. We can make chlorine from salt. Chlorine is used to make drinking water safe by killing harmful bacteria.

FIGURE 1.3.9a: Salt crystals

Sea salt

People have obtained **salt** from seawater and salt lakes for thousands of years. The Romans in Britain trapped seawater in shallow ponds. As the water **evaporated**, salt (**sodium chloride**) crystals were left behind. This method is still used in some countries, such as Australia and India. If the seawater is taken from an area that is not polluted, the crystals of sea salt are pure enough to use in cooking.

1. Give two uses for salt.

2. Name one chemical that is made from salt and explain how it is used.

3. How did the Romans obtain salt in Britain?

4. Why do you think the evaporation method of extracting salt is still used in Australia but not in Britain?

Rock salt

In winter you often see lorries spreading salt and grit on the roads. The salt makes the ice melt and the roads are made safer. This kind of salt is called **rock salt**. The rock salt is mined from the ground and broken into a powder, which makes it easier to spread on icy roads.

Rock salt is not pure salt. It is a natural mixture of salt and insoluble materials like clay. Most rock salt is brown, but it can be yellow or red depending on the clay it is mixed with.

You can purify rock salt yourself. Because the salt is soluble but the sand is not, water can be used to help separate the substances. The insoluble materials can be filtered off using a funnel and filter paper, and the remaining solution evaporated to obtain the salt.

FIGURE 1.3.9b: India is the world's third largest salt producer, after China and the USA.

5. Give two reasons why rock salt and grit are used on icy roads and pavements.

6. What gives rock salt its colour?

7. What is meant by an 'insoluble material'?

8. Describe how you would separate pure salt from rock salt.

Mining salt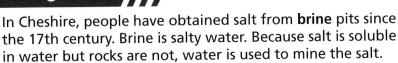

In Cheshire, people have obtained salt from **brine** pits since the 17th century. Brine is salty water. Because salt is soluble in water but rocks are not, water is used to mine the salt. This is called solution mining.

Water is pumped down one of the pipes to dissolve some of the salt underground. The brine is then pumped up and water is evaporated to leave pure, white salt crystals. However, removing the salt from under the ground leaves large holes. The land above can sink into these holes, destroying buildings.

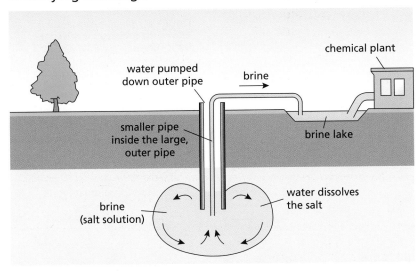

FIGURE 1.3.9c: Using solution mining to extract salt

9. What are the differences between rock salt and the salt extracted by solution mining?

10. Explain, in as much detail as you can, why solution mining is a better source of salt for cooking than rock salt.

11. What are the advantages and disadvantages of solution mining?

Did you know...?

The concentration of salt in the Dead Sea, between Jordan and Israel, is so high that you can float very easily on the water surface. However, only specially adapted microorganisms can survive in it.

Key vocabulary

salt

evaporate

sodium chloride

rock salt

brine

Understanding distillation

We are learning how to:

- Use distillation to separate substances.
- Explain why distillation can purify substances.

Distillation is used in making perfumes, fuels (such as petrol) and alcoholic drinks (such as vodka). It is an important separation process involving heating and cooling.

Heating and cooling

On a cold day water **vapour** from a bath or kettle can **condense** on a cold surface. It cools down and turns back to water. This is what happens in **distillation**. Liquid mixtures can be separated using distillation.

1. Name three substances that are made using distillation.
2. Why does steam turn into liquid water when it touches a window?

FIGURE 1.3.10a: Condensation is one of the processes involved in distillation.

Catching steam

When water boils it is hard to catch all of the water vapour because it mixes into the air. In distillation the vapour is cooled, which allows it to be collected as a liquid.

2000 years ago Greek scientists, known as alchemists, invented a way to distil liquids. A copper 'still' trapped the hot vapour, cooling it and condensing it back to a liquid. It was so successful that the design was on sale until 1860.

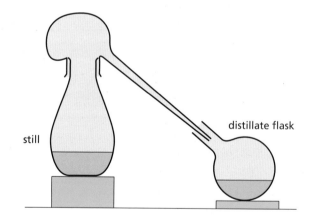

FIGURE 1.3.10b: An early distillation method

The distillation apparatus that we use today is based on the same principle of heating and cooling. The major improvement is the **Liebig condenser**, which is a double glass tube, shown in the apparatus in Figure 1.3.10c. The hot vapour from the boiling liquid flows through the inner tube, while cold water runs though the outer tube. This keeps the inner glass tube cold and condenses most vapours easily. The liquid collected at the end is called the distillate.

104 KS3 Science Book 1: Mixing, Dissolving and Separating

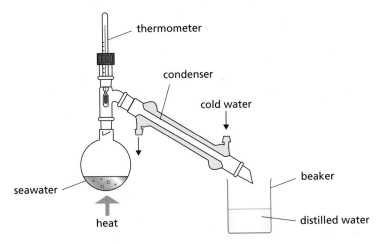

FIGURE 1.3.10c: Distillation apparatus using a Liebig condenser

3. Why is the alchemists' method a better method of separation than just heating a mixture of liquids?

4. Why is the Liebig condenser better than the alchemists' equipment?

5. Explain the safety checks you would use to separate a mixture safely.

Distilling mixtures

There are two changes of state in distillation. First, a liquid is evaporated by heating and then the cooled vapour condensed back to a liquid. When salty water is heated, only the water (solvent) changes state and the salt (solute) is left behind. The water produced is called distilled water.

Different liquids boil at different temperatures – for example, ethanol boils at 78°C and water at 100°C. This means that mixtures of liquids can be separated using distillation. A thermometer at the top of a distillation flask shows the temperature of the vapour being condensed and hence identifies the substance being separated. Distillation is an effective way of **purifying** alcohol or increasing the concentration of alcoholic drinks. It is also useful for separating flammable liquids like petrol and diesel because the vapours never come into direct contact with the flame.

6. Why is distillation a better way to separate salt and water than crystallisation?

7. Why is a thermometer important in distillation?

8. Why is distillation a good method for separating petrol and diesel?

9. Explain how water and ethanol are separated.

> **Did you know…?**
>
> Steam distillation is used to obtain essential oils from plants such as herbs and flowers. The products are used in aromatherapy, flavourings in foods and drinks, and as scents in perfumes, cosmetics and cleaning products.

Key vocabulary

vapour

condense

distillation

Liebig condenser

purify

Applying key ideas

You have now met a number of important ideas in this chapter. This activity gives an opportunity for you to apply them, just as scientists do. Read the text first, then have a go at the tasks. The first few are fairly easy – then they get a bit more challenging.

How hard is your water?

Some people have hard water coming out of their taps. It does not really do them any harm, but it can cause problems. A good way to tell if the water is hard in the area where you live is to look inside your kettle at home – in a hard-water area you will see limescale there. It looks like a creamy coloured fur but if you touch it the 'fur' is actually quite hard. You can also tell if you go to wash your hands – the soap will not form much lather in the water, but will tend to form a scum instead.

FIGURE 1.3.11a: Limescale in a kettle

Water is hard if it has minerals dissolved in it that contain calcium or magnesium. Some dissolved chemicals cause the hardness to be temporary. These chemicals can be removed by boiling and the water becomes soft. Calcium bicarbonate or magnesium bicarbonate cause temporary hardness. However, if there is calcium sulfate or magnesium sulfate in the water the hardness is permanent – boiling the water will not remove it.

As well as forming deposits on the inside of kettles, the chemicals in hard water can also cause harm to central heating systems, dishwashers and washing machines by clogging them up.

Now, you might wonder how these chemicals get into the water. After all, water falls as rain and rain is considered to be soft – it contains few chemicals. Actually, rain water is slightly acidic (this is natural and, unlike the acid rain caused by pollution, not a problem). The way that water becomes hard is by running over chalk or limestone. Limestone, for example, contains calcium carbonate and the rainwater reacts with this.

FIGURE 1.3.11b: A limestone landscape

Not everybody has hard water though, and in a number of areas it is soft, depending on where the water comes from. The water in Birmingham, for example, comes from reservoirs in the Elan Valley in Wales, which is a soft-water area. Soft water lathers with soap quite easily and does not cause deposits. However, it is not all good news for people in soft-water areas – some people prefer the taste of hard water.

Task 1: Is your water hard?

Is the water in your area hard? How do you know? People can often taste the difference between hard water and soft water. How can water taste different?

Task 2: Temporary or permanent?

What sort of problems does hard water cause? How does water become hard? What is the difference between temporary hard water and permanent hard water? What causes that difference?

Task 3: Soluble or insoluble?

Limescale is calcium carbonate. Is it soluble or insoluble? How do you know? How could you demonstrate this? What about calcium bicarbonate?

Task 4: Calcium carbonate

Calcium carbonate is made up from the elements calcium, carbon and oxygen and is written as the chemical 'formula' $CaCO_3$. Which symbols represent each element? How can you tell from this formula that each molecule consists of five atoms?

Task 5: Calcium hydrogen carbonate

Calcium bicarbonate has the formula $Ca(HCO_3)_2$. Which elements does it contain? Why do you think it is called calcium *bi*carbonate? Its modern name is calcium hydrogen carbonate. Why is this more useful?

Finding out what air is made of

We are learning how to:

- Describe the composition of air.
- Separate gases from air.

Your body is very effective at separating things out from air. Air is a mixture of gases and small particles of dust, pollen and pollutants. Hairs and mucus in your nose help in removing dust and pollen.

FIGURE 1.3.12a: Particles in the air can affect our health.

Up in the air

Two **gases** make up 98 per cent of the air – nitrogen (N_2) and oxygen (O_2) are diatomic. Nitrogen is not a very reactive gas – however, oxygen is needed to burn fuels for heat, to make electricity and for car engines to run. When oxygen reacts with other elements, compounds called oxides are formed. Carbon dioxide (CO_2) is produced when carbon compounds are burned in air. Animals breathe out carbon dioxide.

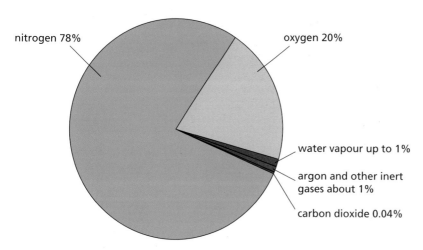

nitrogen 78%

oxygen 20%

water vapour up to 1%

argon and other inert gases about 1%

carbon dioxide 0.04%

FIGURE 1.3.12b: Air is a mixture.

1. Which are the two most common gases in the air?

2. What name is given to any compound formed from a reaction involving oxygen?

3. Give two ways that carbon dioxide can be released into the air.

The particles and gases produced by burning fuels are released into the air. The pollution this causes has short- and long-term health effects. The daily air quality index (DAQI) tells us how much pollution there is in the air and DEFRA, the government body that produces it, provides health advice. This information is vital, for example, for people with lung problems or asthma.

The **air pollution** information is based on hourly results taken from over 120 automatic monitoring systems across the UK. They detect fine particles and gases such as sulfur dioxide and nitrogen oxides that can irritate our lungs.

1	2	3	4	5	6	7	8	9	10
Low			Moderate			High			Very high

FIGURE 1.3.12c: Daily air quality information is displayed using this scale. (DEFRA)

4. Why is the DAQI information important for children with asthma?

5. Why are so many automated monitoring stations in the UK taking readings every hour?

6. Which pollutants are measured and where do they come from?

Boiling cold? ▶▶▶

Most gases have **boiling points** significantly lower than 0 °C. Therefore to separate oxygen from the other gases in air, the distillation process has to take place at very low temperatures. The cryogenic distillation process can produce high-purity gases by pushing the molecules closer. This also means that the gases take up less space. Air is **compressed** and cooled to very low temperatures and the liquid air is allowed to heat up slowly in a fractionating column. Oxygen oils off at −183 °C and the purified gas is stored under pressure.

7. Newborn babies sometimes need to breathe pure oxygen. How is pure oxygen obtained?

8. Why does cryogenic distillation use more energy than normal distillation?

9. Why are gases like oxygen and compressed air stored in pressurised containers?

Did you know…?

'SCUBA' stands for 'self-contained underwater breathing apparatus'. The tanks contain compressed air to allow divers to breathe underwater.

FIGURE 1.3.12d: Scuba divers use tanks of compressed air.

Key vocabulary

gas

air pollution

boiling point

compressed

Exploring chromatography

We are learning how to:

• Use chromatography to separate dyes.

Chromatography is one of the most important separation methods used to identify unknown substances. There are many types of chromatography – some use liquids and some use gases. Chromatography is used by scientists to detect drugs and explosives and to identify dyes and paints.

FIGURE 1.3.13a: Separation by chromatography

Separating colours

Black ink is not just a black colour mixed with water. Black ink is a **mixture** of colours. Filter paper and water can be used to **separate** these colours. This method of separation is called **paper chromatography**. Figure 1.3.13a shows what happens when the colours are separated. Drops of water (solvent) are added to the middle of the paper where the ink spot is placed.

1. What evidence is there that black ink is not pure?

2. What causes the ink to spread across the filter paper?

3. What is chromatography?

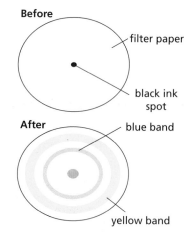

FIGURE 1.3.13b : Separating ink colours

Examples of chromatography

If you cut a section of the filter paper, it can act as a wick. By dipping this wick into water, the liquid is drawn up through the ink and the colour begins to separate all on its own. The resulting pattern of colours is a **chromatogram**.

This method shown in Figure 1.3.13c is called ascending paper chromatography, because the water soaks up from the base carrying the colour spots with it. Some colours move faster than others, which is why the colours separate.

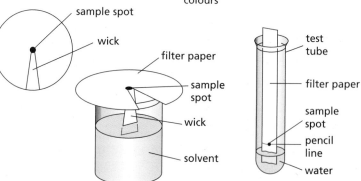

FIGURE 1.3.13c: Other methods of chromatography

You can use chromatography for colourless mixtures, but must develop the chromatogram by spraying the paper with a chemical to make the spots visible, or using an ultraviolet (UV) light to look at the spots.

4. What do we call the pattern of spots on the paper?
5. Why is the line drawn in pencil and not in ink?
6. Why would paper chromatography be no good for separating salt from water?

Making comparisons

If the same conditions are used, the distance that a coloured **dye** travels up the paper is always the same. This is how chromatography can be used to identify unknown substances. The distance that the dye travels from the original spot on the pencil line at the base, compared to the distance travelled by the solvent (solvent front) can be calculated. This is called the R_f or retardation factor.

R_f = distance travelled by dye ÷ distance travelled by solvent

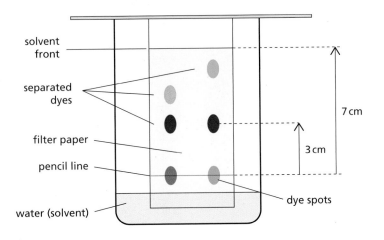

FIGURE 1.3.13e: Chromatogram to compare two dyes

7. Why is the starting line above the water level in the beaker?
8. If two brands of blue dye gave identical spot patterns, what could you conclude?
9. What conclusions can you draw from the chromatogram in Figure 1.3.13e?
10. Some dyes are invisible in normal light and only show up in ultraviolet light. This is how security inks work. How could you use chromatography with a set of security inks?

Did you know...?

You can separate pigments in leaves by chromatography using alcohol as the solvent. Chlorophylls are green pigments that help plants make food via photosynthesis. The yellow pigment separated here is carotene, which is found in carrots and is used as a food colouring (E160a).

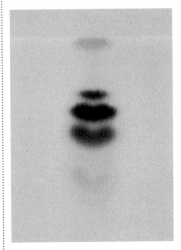

FIGURE 1.3.13d: A plant pigment chromatogram

Key vocabulary

mixture

separate

paper chromatography

chromatogram

dye

Using chromatography

We are learning how to:

- Use chromatography to identify unknown substances.
- Draw conclusions from evidence.

Solving crimes requires scientists to produce high-quality evidence that the court will believe and use to prove guilt or innocence. This means that forensic scientists have to be very careful and repeat their experiments to be sure that their evidence is reliable and accurate.

Major crime solved

A series of burglaries and break-ins at jewellery stores has netted a gang of criminals thousands of pounds. Unfortunately for them they were careless. Fast driving caused accidents that left traces of car paint at several of the crime scenes. One of the robbers left graffiti messages scrawled on the wall, taunting the police to catch them.

The police **forensic** team used the **reliable** technique of **chromatography** to prove that they had arrested the right gang. The paint left at the crime scenes matched that on the criminals' car perfectly.

Chromatograms of the ink from the wall also exactly matched a set of marker pens found in the gang leader's home and provided the **accurate evidence** needed.

1. What did the forensic scientists deduce after making the chromatogram (Figure 1.3.14b) of the car paint found at the scene?

2. How could they make their conclusion more reliable?

3. Is it sufficient evidence to find that the paint on a car matches that at the scene of the crime?

Considering the evidence

Figure 1.3.14c shows a chromatogram using ink from the graffiti messages and ink from four of the pens in a set found in the suspect's home. Consider the evidence from Figure 1.3.14c and answer the following questions.

FIGURE 1.3.14a: The recovered car provided clues.

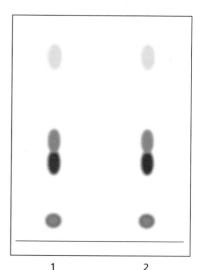

1 Crime scene paint
2 Suspect car paint

FIGURE 1.3.14b: Chromatogram of paint samples

4. What evidence is there that the graffiti ink was a mixture?

5. How does the evidence from the chromatography prove that the graffiti was drawn by the suspect?

6. Explain how the forensics team could make sure that their evidence was reliable and accurate.

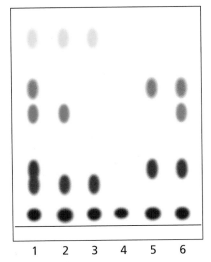

Analysis of ink

1. Crime scene graffiti
2. Suspect pen ink 1a
3. Suspect pen ink 1b
4. Suspect pen ink 2a
5. Suspect pen ink 2b

FIGURE 1.3.14c: Chromatogram of ink samples

Special separation »»»

Samples of DNA gathered at crime scenes can be used to identify and eliminate suspects. The sample is treated with special chemicals and then injected into a gel. When an electric current passes through the gel, the components of the DNA separate and spread, just like the ink on the chromatography paper. This is called electrophoresis. The pattern that the DNA produces is unique to an individual person, like a fingerprint.

Scientists can use DNA 'fingerprints' to find out who you are related to. Your DNA fingerprint contains aspects of the DNA patterns of each of your parents.

7. Explain how DNA fingerprinting is similar to chromatography.

8. What are the differences between chromatography and electrophoresis?

9. What precautions would forensic scientists have to take when gathering and testing DNA evidence?

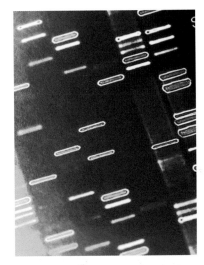

FIGURE 1.3.14d: DNA fingerprinting can help to rule suspects in or out.

Did you know...?

'Chroma' is the Greek word for colour.

Key vocabulary

forensic

reliable

chromatography

accurate

evidence

Finding the best solvent

We are learning how to:

- Choose the best solvent.
- Recognise hazards when using solvents.

Most graffiti artists use spray paints that cannot be washed away with water. To remove spray paint, a different solvent is needed to dissolve the paint.

FIGURE 1.3.15a: Some graffiti has become famous.

Solvent choice

There are many materials like spray paint and oil that are **insoluble** in water, so we need other **solvents** to **dissolve** them. For example, petrol is a very good solvent for oily stains and spray paint, but it is too dangerous to use because it is very flammable. Scientists use **hazard** symbols to highlight the risks of chemicals like solvents, see Table 1.3.15.

1. Why can we not simply wash away graffiti with water?

2. Which solvent would you choose to remove:

   a) nail polish from glass?

   b) ballpoint pen mark from a shirt?

   c) emulsion paint from paint brushes?

3. What hazards are involved when using solvents other than water?

FIGURE 1.3.15b: You can buy special stain removers.

TABLE 1.3.15: Properties of solvents

Solvent	It can dissolve	Hazards
water	sugar, food colours, emulsion paint	none
alcohol (ethanol)	ballpoint pen ink, perfume, herbs, spices	(flammable)
acetone (propanone)	nail polish	(flammable) (harmful/irritant)
white spirit	grease, oil paint	(flammable) (environmental hazard) (health hazard)

Careful choice

Tar stuck to the paint of a car is insoluble in water and difficult to remove. If you try to clean it off with solvents like acetone or white spirit, they will not only remove the tar, but also the paint on the car.

The choice of solvent is very important and must be selected by careful testing and checking. This is important for clothing too. If you use the wrong solvent you could damage the dyes and fabric.

> 4. What are the advantages and disadvantages of using tar and stone chippings on road surfaces?
>
> 5. Why might there be a problem in removing tar from a car?
>
> 6. Explain why there is a need to buy different stain removers when removing stains from clothing.

Clean and smelly

Dry cleaning uses a solvent called tetrachloroethene (C_2Cl_4) instead of water. The clothes are washed in the solvent at 30 °C before being tumbled in warm air (60 °C) to remove it. All the vapours produced are cooled and the condensed solvent is collected. Dry cleaners can recycle nearly 100 per cent of the solvent. This is important because the solvent is classified as highly toxic as well as harmful to the environment.

Alcohols like ethanol can dissolve colours, flavours and odours to make scented products like perfume. Alcohol evaporates easily, which is why perfume or aftershave dries so quickly on your skin. Liquids that evaporate quickly are described as **volatile**. This property allows us to smell substances, but it can also make solvents more dangerous. This is because the substances are more flammable and can enter the lungs quickly when they are vapours.

> 7. How is the dry-cleaning solvent recycled?
>
> 8. Why is it important that dry cleaners recycle as much solvent as possible?
>
> 9. What are the advantages and disadvantages of using a volatile solvent?

Did you know…?

Metals can dissolve in each other to form alloys. Dentists used to use mercury alloys for fillings in teeth. Now most fillings are made from ceramics that avoid the possible harmful effects of metal alloys in your mouth.

FIGURE 1.3.15c: Perfume evaporates easily.

Key vocabulary

insoluble

solvent

dissolve

hazard

volatile

Modelling mixtures and separation

We are learning how to:

- Explain what happens to mass during dissolving.
- Use a circle model to explain dissolving and separation.

Architects use drawings and models to show how new buildings might look. These models are smaller, simpler copies of the real thing. Similarly, we can use a scientific **model** to help explain observations in science.

hydrogen atom, H

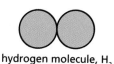

oxygen atom, O

Circle models of substances

A circle model helps us to understand the composition of substances. Each circle represents an **atom**. Different colours represent different atoms. When atoms bond together they form a **molecule**. If a molecule contains different types of atom it is called a compound. If a substance only contains one type of atom or molecule it is **pure**. Mixtures can contain any number or type of different atoms or molecules.

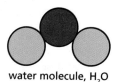

hydrogen molecule, H$_2$

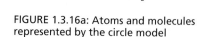

water molecule, H$_2$O

FIGURE 1.3.16a: Atoms and molecules represented by the circle model

1. Why are circle models useful?

2. Draw circle models to represent:

   a) one atom

   b) a molecule consisting of two identical atoms

   c) a molecule consisting of two different atoms

   d) a mixture of different molecules.

Conservation of mass

When salt dissolves in water the crystals seem to disappear, but they are still there, mixed with the water molecules. This can be proved by measuring the mass of the substances. No atoms are lost and no new atoms are added. This is called the **Law of Conservation of Mass**.

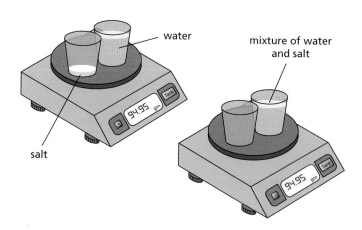

FIGURE 1.3.16b: The mass of water and salt on the first balance is the same as the mass of the mixture on the second balance.

3. What is a solution?

4. How does the mass of a solution show that the solute has not disappeared?

5. Draw diagrams, using the circle model, of a) salt and water, b) salt dissolved in water, to illustrate the Law of Conservation of Mass.

6. The same experiment was repeated, but this time the mass of the solution was lower than the original mass of the salt and water. What might have happened?

Modelling ▶▶▶

In Figure 1.3.16c, showing filtration, the filtrate particles are drawn smaller to show that it can pass through the filter paper. Different colours are used to represent the different substances. The filtrate is a mixture of the smaller particles of two substances.

Models simplify a process to help explain how things work. You do not always need to see all the detail to understand what is happening.

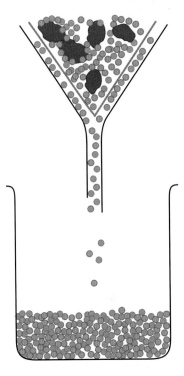

FIGURE 1.3.16c: A model of filtration

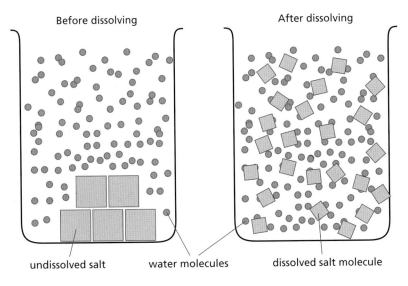

undissolved salt   water molecules   dissolved salt molecule

FIGURE 1.3.16d: This model shows dissolving.

7. In what ways is the model shown in Figure 1.3.16d a good one for dissolving?

8. How could this model be improved?

9. Draw a model to show:

   a) evaporation of water from salt solution

   b) separation of iron filings from sand.

Did you know…?

Chocolate is a mixture. The ratio of cocoa, sugar, milk and fat is carefully controlled and blended so that the chocolate melts in your mouth at the right temperature.

Key vocabulary

model

atom

molecule

pure

Law of Conservation of Mass

Checking your progress

To make good progress in understanding science you need to focus on these ideas and skills.

- Name and draw equipment and explain obvious laboratory risks.
- Select and draw apparatus accurately; explain safety precautions.
- Justify equipment choice and measurements; explain how to reduce risks.

- Use 2D images to represent a range of laboratory equipment.
- Use laboratory equipment safely to gather evidence.
- Record evidence in an effective way.

- Describe how to separate mixtures.
- Select and explain appropriate separation techniques.
- Explain the choice and method of separation using correct terms.

- Describe the process of dissolving and the effect of temperature.
- Describe methods for producing crystals of different sizes.
- Use data to draw conclusions about solubility.

- Understand that seawater is a mixture.
- Explain why most water is not pure, and why this is not necessarily a problem.
- Explain why contaminated water is a problem and identify what can be done about it.

- Identify sources and uses of salt.
- Describe how salt is extracted.
- Recognise advantages and disadvantages of salt extraction methods.

Describe the process of distillation.

Explain the physical processes involved in distillation.

Identify the uses and advantages of distillation.

Describe the composition of air.

Identify sources of air pollution and their impact.

Explain how distillation can be used to separate gases in air.

Identify mixtures using chromatography.

Explain how to separate a mixture using chromatography and interpret chromatograms.

Use chromatograms to explain the composition of mixtures; compare chromatography and DNA analysis.

Explain the idea of a solvent.

Explain mass changes during dissolving; select solvents for different uses.

Use a model to explain dissolving and separation; link the uses of solvents to their properties.

Questions

Questions 1–6

See how well you have understood the ideas in the chapter.

1. What is the chemical symbol for copper? [1]
   a) Pc   **b)** Co   **c)** Cp   **d)** Cu

2. What do we mean by an insoluble material? [1]
   a) It will not dissolve.   **b)** You cannot get it back after it has dissolved.
   c) It dissolves other things well.   **d)** It will dissolve easily.

3. Which are the two most common gases in the air? [2]
   a) helium   **b)** nitrogen   **c)** carbon dioxide   **d)** oxygen

4. Explain two ways in which you could reduce the risks when heating a substance in a test tube. [2]

5. You are trying to get a number of different stains out of your school bag. Give two reasons why you might need to use different stain removers for different stains. [2]

6. Explain how you could safely demonstrate that salt can be recovered from brine (salt water). [4]

Questions 7–12

See how well you can apply the ideas in this chapter to new situations.

7. A good way of separating a mixture of petrol and water is: [1]
   a) distillation   **b)** filtering   **c)** crystallisation   **d)** chromatography

8. Jared likes lots of sugar in his coffee. However, one day he forgets to drink it and tips the cold coffee away. He notices that the sugar has formed a sludge in the bottom of the mug, which he was surprised about because he knew he'd stirred it all in. What is the most likely explanation? [1]
   a) If you leave it, sugar doesn't stay dissolved in a liquid for long.
   b) Sugar is insoluble in coffee.
   c) The sugar had reacted with the coffee.
   d) Not as much sugar could dissolve in the cold coffee as in the hot coffee.

9. A sample of water was heated until it all evaporated and left a thin layer of white crystals behind. This proved that:
   a) it was salt water   **b)** it had something dissolved in it
   c) it was pure water   **d)** it was not fit to drink

10. Sally is trying to purify muddy water by filtering it. Which of these statements is *not* true? [1]

    a) This will remove bits of silt that are floating in the water.

    b) It will remove things that have dissolved in it.

    c) The water will be clearer after filtering.

    d) It will not remove soluble solids.

11. Marcus and Lisa are investigating how well different brands of sugar dissolve in water. Using Table 1.3.18a:

    a) Explain what the results show. [2]

    b) What do they have to do to make sure it is a fair test? [2]

TABLE 1.3.18a

Brand of sugar	Mass used (g)	Volume water (cm³)	How many stirs to dissolve it all
Acme	10	100	14
Bonzo	10	100	8
Carefree	20	200	12
Delightful	20	200	7

12. Maisie has three different makes of black felt-tip pen and wonders if the manufacturers made their inks in the same way. Describe an experiment she could conduct to see if the three makes of ink all contain the same colours. [4]

Questions 13–14

See how well you can understand and explain new ideas and evidence.

13. James is conducting an investigation into how well a sweetener dissolves. What does the data in Table 1.3.18b tell you about the solubility of the sweetener at different temperatures? [2]

TABLE 1.3.18b

Temperature of water (°C)	20	30	40	50
Mass of sweetener that can dissolve (g)	120	135	155	180

14. A vet has been asked to find out if any of four horses, A, B, C and D, have been drugged. She takes urine samples from the four horses and arranges for a lab to prepare a chromatogram to test the samples. What do the results show? [4]

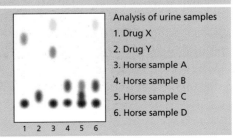

Analysis of urine samples
1. Drug X
2. Drug Y
3. Horse sample A
4. Horse sample B
5. Horse sample C
6. Horse sample D

FIGURE 1.3.18: Chromatogram of horse urine samples

CHEMISTRY
Elements, Compounds and Reactions

Ideas you have met before

Materials

Different substances are made of different materials. Materials have different properties; some are harder than others, some are shinier and some are heavier.

Glass, for example, is a very different material from plastic or metal.

Metals

Metals are shiny solids that we use for many different things, such as making cars, computers, bridges and so on.

Metals are good electrical conductors, which is why we use them to make wires and circuits.

States of matter: solids, liquids and gases

Most materials can be classified into one of three groups: solids, liquids and gases.

Ice, water and steam are three states of matter of the same substance. We can convert materials from solids to liquids and gases by heating or cooling them.

Physical and chemical changes

Melting ice is reversible. We can put it into a freezer and produce ice again. This is a physical change.

Some changes are not reversible. These are are called chemical changes. Making toast is a chemical change; you can't change it back into the bread it was made from.

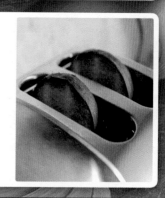

In this chapter you will find out ▷▷

Elements and atoms

- Since ancient times materials have been described in terms of the chemical elements they contain.

- Ideas about elements have changed over time.

- Each element is unique, with its own properties.

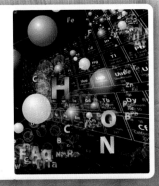

The Periodic Table

- The chemist's dictionary is called the Periodic Table.

- The ingredients of the entire Universe are listed in one place.

Using simple models

- Chemists can represent the building blocks of all materials using simple circle models and symbols.

- Chemical models and symbols help us understand how elements join and react together to make new materials.

Reactions

- Chemical elements can join together in many ways to produce an amazing range of different substances.

- We can make new materials in chemical reactions and then use them to do many different things.

Finding elements and building the Periodic Table

We are learning how to:

- Identify where and how different elements were found.
- Recognise differences between elements.
- Recognise that the Periodic Table has changed over time.

The Sun is about 72 per cent hydrogen and 26 per cent helium, with traces of other elements. Unlike in the Sun, most elements on Earth are found as compounds rather than as elements.

Building blocks of the Universe

Ancient Greek philosophers believed that four **elements** (earth, air, fire and water) made up all matter. Some of the earliest elements to be identified were those that occur naturally but are often trapped in rocks. These are **native elements** and include gold, silver, copper, lead, tin, carbon and sulfur.

FIGURE 1.4.2a: The same elements occur in the Sun and on Earth.

1. Explain the difference between elements found in the Sun and on the Earth.

2. How might the ancient Greeks have classified:

   a) rain?   b) soil?   c) the Sun?   d) steam?

3. Why are gold and sulfur called 'native' elements?

Finding elements

Element	Abundance (% by weight)
aluminium	8.2
calcium	4.2
hydrogen	0.1
iron	5.6
magnesium	2.3
oxygen	46.1
potassium	2.1
silicon	28.2
sodium	2.4
titanium	0.6

FIGURE 1.4.2b: Native gold

TABLE 1.4.2: The 10 most abundant elements in the Earth's crust

The next series of metals to be identified were those released from their **compounds** by heating, such as iron. Their identification has depended on both accidental discoveries and methodical experiments.

Hennig Brand discovered phosphorus by accident in 1669 while heating and purifying urine to find the 'magic substance' that would change metals to gold. In the early 1800s Humphrey Davy used electricity-based experiments to isolate potassium and sodium.

4 .2

4. Why are the ideas of chemists like Davy more reliable than those of the ancient Greek philosophers?

5. Why do you think it took so long for metals like sodium and potassium to be discovered?

The Periodic Table

John Dalton first defined elements in terms of their different **atoms** in 1807. Dmitri Mendeleev was the first to publish a **Periodic Table** of elements, in 1869, basing it on the results of experiments to compare the differences and similarities between them. His table contained 64 elements with gaps for new elements that had not yet been discovered. All these gaps have now been filled and new elements have only been produced by smashing lighter atoms into each other. These man-made elements contain very heavy, unstable atoms that often exist for less than a second before breaking down.

Did you know...?

New elements are all temporarily named using the number corresponding to their position in the Periodic Table. Element 116 was called un-un-hexium before it was named livermorium and given the symbol Lv.

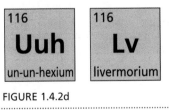

FIGURE 1.4.2d

Reihen	Gruppe I. R^2O	Gruppe II. RO	Gruppe III. R^2O^3	Gruppe IV. RH^4 RO^2	Gruppe V. RH^3 R^2O^5	Gruppe VI. RH^2 RO^3	Gruppe VII. RH R^2O^7	Gruppe VIII. RO^4
1	H=1							
2	Li=7	Be=9,4	B=11	C=12	N=14	O=16	F=19	
3	Na=23	Mg=24	Al=27,3	Si=28	P=31	S=32	Cl=35,5	
4	K=39	Ca=40	—=44	Ti=48	V=51	Cr=52	Mn=55	Fe=56, Co=59, Ni=59, Cu=63.
5	(Cu=63)	Zn=65	—=68	—=72	As=75	Se=78	Br=80	
6	Rb=85	Sr=87	?Yt=88	Zr=90	Nb=94	Mo=96	—=100	Ru=104, Rh=104, Pd=106, Ag=108.
7	(Ag=108)	Cd=112	In=113	Sn=118	Sb=122	Te=125	J=127	
8	Cs=133	Ba=137	?Di=138	?Ce=140	—	—	—	— — — —
9	(—)	—	—	—	—	—	—	
10	—	—	?Er=178	?La=180	Ta=182	W=184	—	Os=195, Ir=197, Pt=198, Au=199.
11	(Au=199)	Hg=200	Tl=204	Pb=207	Bi=208	—	—	— — — —
12	—	—	—	Th=231	—	U=240	—	

FIGURE 1.4.2c: The Periodic Table showing gaps Mendeleev identified

6. How do you know that there are no more naturally occurring elements to be discovered?

7. Why might some people say that artificial elements are not true elements?

8. Suggest what temporary name would be given to element 115.

Key vocabulary

element

native element

compound

atom

Periodic Table

Looking at the Periodic Table of elements

We are learning how to:

- Navigate the Periodic Table and identify some of the elements.
- Identify features of the Periodic Table and describe how it is organised.
- Explain why the Periodic Table is useful.

The Periodic Table lists all the known chemical elements in the Universe. The patterns and trends in the arrangement help chemists explain and predict the behaviour, properties and reactions of all the elements.

Periods and groups

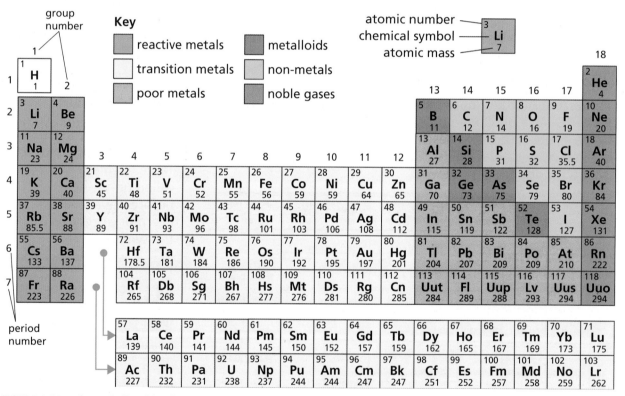

FIGURE 1.4.3a: The Periodic Table of elements

The **Periodic Table** is arranged in rows called **periods** and columns called **groups**. Groups are families of elements with similar properties. Group 1 contains the alkali metals, which all react quickly with water. The halogens are in Group 17; they are good at killing bacteria. The noble gases in Group 18 are all unreactive gases. These characteristics are called chemical trends and patterns.

Another pattern to recognise is that metals are on the left and non-metals, except hydrogen, are on the right. In between are metalloids, which have some of the properties of metals but not all.

1. How many groups make up the Periodic Table?

2. Name three families of elements.

3. The Periodic Table is split into three main types of elements – what are they called?

Atomic number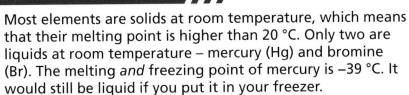

Each element has unique number, called its **atomic number**. This number increases from left to right across each period. For example, hydrogen (H) is number 1, lithium (Li) is number 3, carbon (C) is number 6 and neon (Ne) is number 10. This is an important pattern in the Periodic Table.

4. Describe how the elements are arranged in the Periodic Table.

5. Use the Periodic Table to answer these questions:

    a) In which group would you find carbon (C)?

    b) In which period would you find magnesium (Mg)?

Melting and boiling

Most elements are solids at room temperature, which means that their melting point is higher than 20 °C. Only two are liquids at room temperature – mercury (Hg) and bromine (Br). The melting *and* freezing point of mercury is −39 °C. It would still be liquid if you put it in your freezer.

Oxygen (O) is a gas at room temperature, which means that it has a boiling point below 20 °C. To turn oxygen into a liquid you would have to cool it to below −183 °C.

6. Look at Table 1.4.3 showing the melting and boiling points of five different elements (V–Z).

    a) Which one has the lowest boiling point?

    b) Which one has the highest melting point?

    c) Which one is liquid at room temperature?

7. If the temperature dropped to below room temperature which element would freeze first? Explain your choice.

8. What is the number of the element with a boiling point lower than oxygen?

Did you know…?

Mercury is sometimes called quicksilver and is the only metal that is liquid at room temperature. It was named after the Roman messenger of the gods. Its symbol Hg is derived from the Greek word hydrargyros, which means silver water.

TABLE 1.4.3: Trends and patterns in melting and boiling points tell us about the physical state of elements at different temperatures.

	Melting point (°C)	Boiling point (°C)
V	−210	−196
W	−7	59
X	328	1750
Y	1064	2856
Z	115	445

Key vocabulary

Periodic Table

period

group

atomic number

Understanding elements and atoms

We are learning how to:

- Interpret chemical symbols.
- Explain what is meant by 'element' and 'atom'.
- Work out the composition of different substances based on their names.

Everything is made of atoms. Every material in the Universe is composed of different combinations of just 92 naturally occurring atoms. Each one of these atoms is represented by a unique symbol. These form the chemical alphabet of a common language used by chemists all over the world.

Elements and atomic numbers

Atoms are so small that you would find 5 000 000 000 000 000 000 000 (5×10^{21}) of them in a single drop of water. Each element has a symbol and two numbers that help to identify it. No two elements have the same symbol or **atomic number** because they contain different types of atoms.

1. Why are elements given unique symbols?

2. Why can two different elements not have the same atomic number?

FIGURE 1.4.4a: All elements have a symbol, an atomic number and an atomic mass. This represents the element gold.

Atomic mass

Elements also have an **atomic mass** number. This number allows us to compare elements and calculate how much we need to make a product.

TABLE 1.4.4: Atomic number and atomic mass of some elements

Element	Symbol	Atomic number	Atomic mass
hydrogen	H	1	1
gold	Au	79	197
carbon	C	6	12
calcium	Ca	20	40
lead	Pb	82	207
krypton	Kr	36	84

FIGURE 1.4.4b: Pure gold is made up of only one type of atom.

Pure gold is an element because it is made up of one type of atom. It has the symbol Au, an atomic number of 79 and an atomic mass of 197. This means that gold is the 79th element in the Periodic Table and that its atoms are 197 times heavier than hydrogen atoms.

3. Look at the Periodic Table in Figure 1.4.3a to answer these questions.

   a) What is the atomic number of carbon?

   b) What is the atomic mass of calcium?

   c) How many times greater is the mass of one krypton atom than one carbon atom?

4. What are elements made of?

Useful compounds

When different elements combine they form new substances called **compounds**. Each compound has its own unique properties. By understanding how different elements combine and how compounds react together, chemists can design new substances, such as medicines and materials, useful in everyday life. For example, indium oxide and tin oxide compounds can be combined to produce a transparent electrical conductor. Indium-tin oxide is used in flat-screens and touch screens for televisions, computers and mobile phones. Indium is a rare and therefore expensive element.

5. What is the difference between an element and a compound?

6. Look at the structure of the drug cisplatin in Figure 1.4.4d. Identify the names of the different elements and how many atoms of each the compound contains.

7. Using the correct chemical terms, explain what indium tin oxide is.

8. Use the Periodic Table in Figure 1.4.3a to find out which element has atoms that are:

   a) 12 times heavier than hydrogen

   b) twice as heavy as calcium

   c) positioned 26th in the Periodic Table

   d) twice as heavy as neon and with an atomic number half that of zirconium.

Did you know…?

Atoms are too small to be seen through ordinary microscopes. Special electron microscopes are needed to show us how atoms are arranged inside materials.

FIGURE 1.4.4c: You could line up about 78 million copper atoms across a 1p coin!

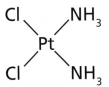

FIGURE 1.4.4d: The compound cisplatin is a drug used to treat cancer.

Key vocabulary

atom

atomic number

atomic mass

compound

Understanding metals

We are learning how to:

- Recognise the properties and uses of metals.
- Identify differences between metals.

Some metals, like gold, occur naturally; some, like sodium, are so tightly bound into their compounds that electricity is needed to extract them; some, like uranium, are radioactive. Only three are magnetic and only one is a liquid at room temperature.

Metals and their properties

We use metals for building because they are strong and for making jewellery because they are shiny and attractive. *Strong* and *shiny* are two properties of metals.

We use metals in electrical circuits because they all conduct electricity and are **ductile**, which means they can be stretched into wires. Metals are **malleable**, meaning that they can be bent, rolled into sheets and shaped without them breaking. Most metals make a ringing noise when hit, which means that they are **sonorous**. They are also very good **conductors** of heat and most have high melting points.

FIGURE 1.4.5a: Cars rely on different properties of metals.

1. List the properties common to most metals.

2. Which properties of metals are most important for making:

   a) saucepans?

   b) water pipes?

   c) drinks cans?

Transition metals and alloys

Iron and copper are 'transition metals', which are found in the centre of the Periodic Table (Figure 1.4.3a). Iron is a very strong, grey metal which makes it useful as a structural material. Copper is an orange coloured metal that is more malleable and ductile than iron. It is used in electrical circuits, wires and water pipes. Water pipes were made of lead until it was found to be harmful to living things. Unlike iron, copper does not rust.

FIGURE 1.4.5b: Metals have many useful properties, so they have wide-ranging uses.

Iron, cobalt and nickel are the only magnetic metals.

Metals can be mixed together to form alloys. Alloys have different properties compared to the metals that they are made from, which sometimes makes them more useful. Mixing zinc with copper forms a harder metal alloy called brass, which is used for making door locks and heating pipes. Stainless steel is an alloy of iron – adding different metals, like chromium, to the iron makes it stronger, shinier and less likely to rust.

3. Name two metals and the property that makes them different from most other metals.

4. Identify three differences between the properties of

   a) iron and copper   **b)** iron and stainless steel

5. What is an alloy and why are they used instead of pure metals?

FIGURE 1.4.5c: Iron bridge

Group 1 metals

The alkali metals, such as sodium, are found in Group 1 of the Periodic Table. They are softer than other metals and can be cut with a knife. They react quickly with air and water and are stored in oil to stop the oxygen and moisture in the air reaching the metal surface and reacting with it.

FIGURE 1.4.5d: Sodium is a soft reactive metal.

6. Look at Table 1.4.5. Why would it not be a good idea to make:

   a) a necklace from sodium?

   b) electrical wiring in a house from rhodium?

7. If platinum costs £28 per gram and rhodium £21 per gram, is it better to own a 42 g rhodium ring or a 31 g platinum necklace?

TABLE 1.4.5

Top five metal electrical conductors	Top five most expensive metals
1 silver	1 technetium
2 copper	2 platinum
3 gold	3 gold
4 aluminium	4 rhodium
5 rhodium	5 rhenium

Did you know…?

Mobile phones contain more than 10 different metals including some of the rarest. The battery contains copper, cobalt, zinc and nickel. The circuit board and touchscreen may contain copper, gold, arsenic, cadmium, lead, nickel, silver, zinc, mercury, indium and tantalum.

Key vocabulary

ductile

malleable

sonorous

conductor

Understanding non-metals

We are learning how to:

- Identify the uses of common non-metals.
- Describe the properties of non-metals.

Non-metals are neither strong nor shiny and most are unreactive gases at room temperature. They have lower densities than metals and are very poor conductors. Some have special properties and are vital to life.

Meet the halogens

The **halogens** are important elements because they form compounds with metals called **salts**. Fluoride salts are added to water and toothpaste to help keep teeth healthy. Chloride salts like sodium chloride are added to food to give a salty taste. Chlorine compounds are used to kill bacteria in water and swimming pools.

Bromine is a liquid at room temperature but it quickly evaporates to a brown-orange gas. Like chlorine it has a very strong, sharp smell and is very harmful if breathed in or on contact with the skin.

Iodine doesn't melt. When it is heated, the dark grey crystals turn straight into a purple vapour. Iodine is a good antiseptic and is also essential in your diet. Lack of iodine causes a swelling called a goitre in the thyroid gland in your neck.

FIGURE 1.4.6a: Bromine and iodine are colourful elements.

1. How are the properties of metals and non-metals different?

2. What is the name given to the compounds that halogens form with metals?

3. Why is iodine an unusual element?

Sulfur and its uses

Sulfur is a bright yellow solid that can be crushed into a powder and melted. As an element, it exists in different forms called *allotropes*. When sulfur burns it forms a choking, toxic gas called sulfur dioxide that attacks the throat and lungs.

Sulfur has many important uses including making rubber for car tyres, gunpowder and a very important chemical called sulfuric acid. Sulfur is also essential to life and humans take in about one gram daily.

FIGURE 1.4.6b: Sulfur and sulfur dioxide are produced from volcanoes.

4. How is sulfur both helpful and harmful to us?

5. What is an 'allotrope'?

6. Why should you be worried if someone starts to burn sulfur near to you?

7. Explain why sulfur is considered to be a non-metal.

Noble gases 》》》

The **noble gases** are colourless and odourless. Helium is less dense than air – helium balloons float. All the noble gases are **inert** because their atoms are unreactive. This property makes them very useful because they can be used to stop other elements and compounds reacting with each other. For example, argon gas is used to fill light bulbs so that the metal filament lasts longer.

Neon lights are made by passing electricity through glass tubes containing noble gases. Neon glows red and argon glows blue.

Halogen bulbs or tungsten–halogen lamps contain very small amounts of bromine and iodine gas. These bulbs appear brighter because the halogens allow higher temperatures to be used without shortening the lifespan of the filament inside.

FIGURE 1.4.6c: Helium balloons float because helium is less dense than air.

8. How are the noble gases different from other non-metals?

9. Use the information here and the Periodic Table in Figure 1.4.3a to compare:

   a) helium and chlorine   b) iodine and chlorine

   c) bromine and argon   d) metals and non-metals.

Did you know...?

Chlorine is named after Greek word for green, 'chloros', because of the colour of the gas. It was used as a weapon in World War 1, poisoning thousands on the battlefield.

FIGURE 1.4.6d: The noble gases are used in light bulbs.

Key vocabulary

halogen

salt

noble gas

inert

Identifying metalloids

We are learning how to:

- Describe semi-metals and their properties.
- Identify some common uses of semi-metals.

The **metalloids**, or **semi-metals**, are a series of elements that lie between the metals and non-metals and show properties of the elements on both sides of the Periodic Table. These elements have special properties we rely on as we develop faster and smaller technology.

Arsenic – a Victorian medicine

The semi-metal arsenic is found in compounds in the Earth's crust, in the soil, in water and in the air. Arsenic can be released from cigarette smoke if its compounds were sprayed onto the tobacco plant to reduce insect activity. Shellfish, such as prawns, absorb and accumulate arsenic from polluted water which can be transferred to us or their predators when they are eaten. Although we now know arsenic to be highly poisonous above a certain – very small – dose, Charles Dickens was one of many Victorians to take potassium arsenate as a 'cure-all' tonic. Arsenic combined with gallium is used in electronic circuits.

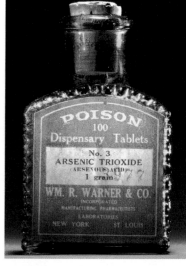

FIGURE 1.4.7a: Arsenic was once used in small doses as medicine.

1. List the useful and the harmful properties of arsenic and its compounds.

2. Describe different ways that arsenic might get into water and pollute it.

Silicon – from chips to crystals

A vast number of compounds that have very different properties and uses are formed with silicon. The tiles for the Space Shuttle were made of a special silicon-based material because they needed to withstand temperatures above 3000 °C.

Silicon chips are miniature electrical circuits etched onto the surface of small, wafer thin pieces of pure silicon. This works because silicon is a **semiconductor**. It conducts electricity, but not as well as a metal or graphite.

The majority of rocks found on Earth contain silicon oxide compounds. As a result, beach sand is our main source of silicon dioxide (SiO_2), which is very useful in producing building materials such as cement, concrete, glass and steel.

When exposed to the heat and pressure inside the Earth, silicon oxides and silicates form attractive **crystalline** solids that we call gemstones.

FIGURE 1.4.7b: Most gemstones contain silicon compounds.

3. Using examples, explain why silicon and its compounds are really useful to us.

4. Explain why silicon should be expected to have similar properties to carbon.

Radioactive elements

Polonium is named after Poland, the birthplace of Marie Curie who discovered the element with her husband Pierre while investigating uranium oxide. Like uranium, polonium is **radioactive**. This means that it releases very dangerous, invisible radiation for possibly long periods of time. Radiation causes damage to living things and can cause cancer. The dangers of radiation were not known when polonium was discovered in 1898 and Marie died, aged 66, of an illness related to her long-term exposure to radioactive substances.

The story of polonium is key to the development of the modern Periodic Table and our understanding of radioactive elements. Radioactive elements are used to produce energy in nuclear power stations and have a variety of uses in medicine, including in the identification and treatment of cancers.

FIGURE 1.4.7c: Marie Curie in her laboratory

Key vocabulary

metalloid

semi-metal

semiconductor

crystalline

radioactive

5. Name two radioactive elements and state why they are dangerous.

6. How can radioactive elements be useful to us?

Discovering the origin of metals

We are learning how to:

- Recognise that metals have to be extracted from ores.
- Evaluate the impact of extracting metals from the Earth.

There are few metals that can simply be dug up from the ground. Metals exist as compounds locked into rocks and in soluble salt compounds in the oceans.

Native metals

Native metals like gold can be found as the pure metal trapped within rocks and river beds. Gold can be separated from rock and sand by panning in water because the heavier gold sinks to the bottom. The Californian Gold Rush (1848–1855) attracted around 300 000 people from all over the world to find their fortune.

1. What does the term 'native metal' mean? Name one example.

2. What is gold removed from during panning?

FIGURE 1.4.8a: Gold miners processing ore in California; a few became wealthy but most made little.

Extracting metals

Most metals exist as compounds trapped inside different rocks or minerals. When there is enough metal to makes its **extraction** worthwhile, these rocks are called **ores**. Bauxite contains 50–70 per cent aluminium oxide. Aluminium is extracted from its ore using electricity – this process is called **electrolysis**.

Copper can sometimes be found as the native metal, but most copper is extracted from ores like malachite, azurite and chalcopyrite which have distinctive green and blue colours.

Iron ore (haematite) is one of the most heavily mined substances in the world. Both iron and copper are extracted from their ores by heating them with carbon. This process is called **reduction**.

3. What is a metal ore?

4. Name two processes used to extract metals from their ores.

5. Why do we need to extract metals from ores?

FIGURE 1.4.8b: Native copper trapped in rock.

Metals ores are mined all over the world. Australia has the world's biggest open gold mine. The Kalgoorlie Super Pit is three kilometres long, a kilometre and a half wide and 400 metres deep.

FIGURE 1.4.8c: The Kalgoorlie Super Pit

Tantalum is a heat-resistant metal used in mobile phones, game consoles and computer chips. It is mined from an ore called coltan in, among other places, the Democratic Republic of the Congo (DRC). The war over mining and mineral wealth in the DRC has caused farmers to be forced from their land; children to abandon school to hunt for the precious metal; destruction of the environment; and the death of endangered species such as the lowland gorilla. It has been estimated that between 1998 and 2002 the war over coltan cost nearly three million lives.

> **Did you know…?**
>
> There are millions of tonnes of gold dissolved in the sea. If you could recover this gold you would be very rich. Unfortunately the cost of extracting the gold would be more than it is worth.

6. What is the impact on local people of mining coltan?

7. List the advantages and disadvantages of living near an open mine.

8. Suggest differences between mining conditions in the DRC and in the Australian Super Pit, Kalgoorlie.

9. Suggest why people in the UK should be concerned about what their mobile phones and computers are made from.

Key vocabulary

extraction

ore

electrolysis

reduction

Choosing elements for a purpose

We are learning how to:

- Recognise elements and their differences from physical data.
- Use data and the properties of elements to choose suitable materials.

It is important that scientists can make good observations and interpret information. They use their knowledge to explain what the patterns in data mean. Companies use scientists to find out the best materials to use in new products.

Gallium

Gallium is a non-toxic, shiny metallic element in Group 3, underneath aluminium. It melts in your hand because it has a low **melting point** of 29.8 °C. It also shatters like glass if you hit it and it will combine with metals like aluminium by forming alloys with them.

1. Why would gallium be a poor choice for a drink can?
2. Which properties are important in a drink can?

FIGURE 1.4.9a: Gallium is a metal that melts in your hand.

Problems with metals

Some elements are **toxic** (poisonous) to humans, making us ill and even causing death over a certain amount. Metals like lead and mercury are stored in our bodies over time and are linked to problems with the brain and the nervous system. They were used in the past in paints, dental fillings, water pipes and make-up, until we realised how harmful they were.

The reactions of different metals with air and water vary. Some, like gold, don't react with either. Group 1 elements like potassium react quickly with oxygen in air and violently with water. Aluminium reacts quickly with oxygen in air but the reaction forms a protective oxide layer over the metal, which stops it reacting with anything else. Iron reacts slowly with air and water forming rust, which weakens the metal.

3. Give two reasons why neither lead nor mercury would be good materials for making a drink can.
4. Compare the use of gold and iron as materials for making drink cans.

FIGURE 1.4.9b: Drink cans are made of metal.

Did you know…?

Gold jewellery and silver coins are 'fake'! Nine carat gold is less than half gold, because pure gold is too soft. 50p coins are 75 per cent copper and 25 per cent nickel, which is cheaper and harder than silver.

Density helps us understand how heavy substances are, compared to their volume. The higher the density, the heavier the substance would be. If blocks of lead and aluminium of the same size were placed on the balance, the lead block would be heavier because it has a higher density.

Sometimes the best material for the job is not chosen because it is too expensive. Silver is a better conductor of electricity than copper, but is only used in specialist items such as satellites.

Table 1.4.9 gives data about the properties of 10 elements, which tell us a lot about their appearance and behaviour.

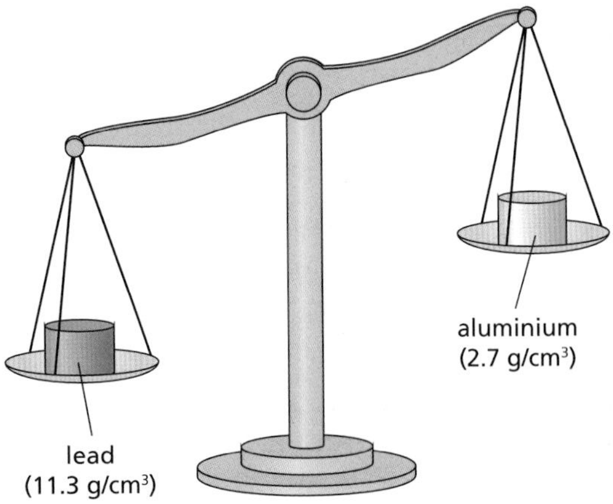

aluminium (2.7 g/cm³)

lead (11.3 g/cm³)

FIGURE 1.4.9c: If different metals are each made into blocks of the same size, the denser one would be heavier.

5. Which of the elements in the table are not solids at room temperature (20 °C)?

TABLE 1.4.9: Properties of some common elements

Element	Conducts heat	Conducts electricity	Melting point (°C)	Density (g/cm³)	Cost pure (£/g)	Other properties
graphite	not well	yes	3730	2.25	1.50	brittle
helium	no	no	−270	0.15	3.30	inert
lead	yes	yes	327	11.30	1.60	poisonous
aluminium	yes	yes	660	2.70	9.90	protective layer
hydrogen	no	no	−259	0.07	7.60	flammable
silver	yes	very well	961	10.50	75.60	shiny
gallium	yes	yes	30	5.91	3.20	poisonous
sodium	yes	yes	98	0.97	15.80	very reactive
iron	yes	yes	1535	7.86	4.50	rusts
copper	yes	very well	1083	8.92	6.20	non-toxic

6. a) Give one reason for and one against using graphite to make a drink can.

b) Choose the material you think would be most suitable for making drink cans and explain why.

c) Rank the elements in Table 1.4.9 in order from most suitable to least suitable for a drink can.

Key vocabulary

melting point

toxic

density

Applying key ideas

You have now met a number of important ideas in this chapter. This activity gives an opportunity for you to apply them, just as scientists do. Read the text first, then have a go at the tasks. The first few are fairly easy – then they get a bit more challenging.

How tinny is a tablet?

If we call something 'tinny', we're probably not being very kind about it. If you said that someone's car was a bit tinny, they might not be very pleased because it implies it isn't very well made. Tin, however, is very important to us, and has been for thousands of years. It is a silvery metal that is both malleable and ductile.

Copper has been used for thousands of years to make tools, coins, weapons and decorations. It isn't very hard, but it can be alloyed with tin to make bronze, which is much harder. Roman officers had swords made of bronze, and for many decades bronze was used to make ships' propellers, until it was replaced by stainless steel.

For thousands of years one of the main sources of tin was Cornwall. At its height, in the 19th century, the Cornish tin mining industry produced 10 000 tonnes of tin a year, from cassiterite, or tin ore, SnO_2. The ore is crushed, washed, roasted to remove sulfur and arsenic as oxides, and then heated strongly with coal to produce pure tin.

Tin doesn't easily react with oxygen and is used to coat other metals such as steel. A tin can isn't made of tin but of steel coated with tin.

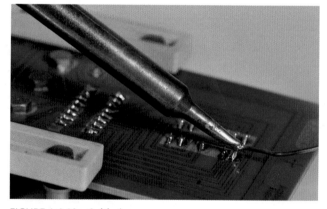

FIGURE 1.4.10a: Soldering

Tin has a low melting point compared with other metals and is used to make solder, which is used to join electrical components together so that they make effective and permanent connections.

When heated, the solder melts and flows onto the contacts; it then sets hard, fixing the component into the circuit.

Every tablet computer contains thousands of soldered joints and, therefore, several grams of tin. In fact, it's one of the most common metals used in their manufacture.

FIGURE 1.4.10b: Tablet

Task 1: Exploring properties

What does tin look like? Explain in simple terms what is meant by tin being silvery, malleable and ductile.

Task 2: Thinking about alloys

What is meant by an alloy? What has to happen to metals to turn them into an alloy? How did tin make copper more useful? Copper and tin are not often mined in the same area. Why did the production of bronze depend upon trade?

Task 3: Applications at sea

What sometimes happens to metals in seawater? What do you think it was about bronze that made it suitable for making ships' propellers?

Task 4: Applications in electronic devices

What is it about tin that makes it suitable for use as solder?

Task 5: Finding its family

Find tin in the Periodic Table. What are its neighbours in that group? Research their properties and find out what they have in common with tin.

Task 6: Thinking about its ore

What is an ore? The formula of tin ore is SnO_2. What does this formula tell you? Tin ore is roasted to drive off sulfur and arsenic as oxides. What would the names of the compounds formed be? Why would this be a dangerous process?

Combining elements

We are learning how to:

- Explain what is meant by a compound.
- Recognise how compounds are formed and named.
- Interpret the ratio of atoms and formula of compounds.

Elements can combine together to form new compounds with different properties from the original elements. This is why there is such a variety of substances made from only 92 naturally occurring elements.

Using models

The building bricks in Figure 1.4.11a can be joined together to create many different structures. Each structure has a different number or type of bricks joined in a particular way. There are only certain ways in which the bricks can join. This is a good **model** to explain how atoms join to create different compounds.

1. What is the difference between elements and compounds?

2. How does the house model help explain why there are many different materials but relatively few elements?

FIGURE 1.4.11a: Bricks can be combined to build a house just as elements combine to form compounds.

Chemical formulas

Chemistry is about the way that atoms combine in compounds. We use the chemical symbols of the elements to write the chemical **formula** of the compound, which represents the **ratio** of atoms in each unit of the compound – as illustrated in Figure 1.4.11b.

1 red brick and 1 blue brick, the ratio is 1:1
This model could represent one unit of the compound copper oxide, CuO

2 red bricks and 1 green brick, the ratio is 2:1
This model could represent one molecule of the compound carbon dioxide, CO_2

FIGURE 1.4.11b: Models showing the ratios of atoms in two compounds.

The name of the compound sometimes gives a clue to the ratio as well as the elements that make it. When oxygen forms a compound its name changes to oxide. When there are two oxygen atoms in the compound, its name changes to dioxide. Other non-metals also change their names.

3. How do the names of compounds help us understand what they are made of?

4. What is the difference between oxygen and an oxide?

5. Use chemical formulas and models to explain how carbon monoxide and carbon dioxide are different.

6. Name the compounds formed when the following elements react together:

   a) potassium and chlorine

   b) nickel and oxygen

   c) lead and iodine

TABLE 1.4.11

Element name	sulfur	chlorine
Compound name	sulfide	chloride
Example	dihydrogen sulfide	sodium chloride
Chemical formula	H_2S	NaCl
Atom ratio	2:1	1:1

Did you know…?

If you cool oxygen to below −183° C, it condenses to a blue magnetic liquid. It is used as fuel for rocket engines such as those that power the Space Shuttle.

Oxides of lead >>>

Lead is a dense metal and oxygen is an invisible gas. When these two elements are chemically combined they make lead oxide. There are three differently coloured lead oxides.

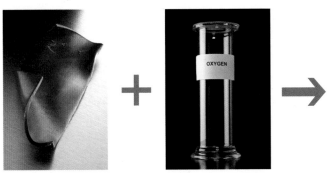 

FIGURE 1.4.11c: Lead reacts with oxygen to form a lead oxide.

7. Calculate the ratios of atoms and give the names of the following compounds:

   a) SO_2 b) NO   c) Al_2O_3 d) CCl_4

8. Suggest a formula for silicon dioxide and explain your choice.

9. Glucose has the chemical formula $C_6H_{12}O_6$. Explain what glucose is made of and give the simplest ratio of its atoms.

Key vocabulary

model

ratio

formula

Using models to understand chemistry

We are learning how to:

- Use a simple model to show the differences between atoms and molecules.
- Use models to represent compounds.

Chemists use chemical symbols and formulas as a shorthand way of explaining chemical structures and reactions, but these can be very complex and hard to understand. As a starting point, atoms can be represented as simple, circular particles to help us explain how they combine and react together.

FIGURE 1.4.12a: Models can aid our understanding in chemistry.

Representing atoms and molecules

Atoms are the smallest parts of all substances. We can represent an atom in three ways: writing its **element** name, writing a symbol or by drawing a simple circular particle.

calcium, Ca   oxygen, O   chlorine, Cl

sodium, Na   hydrogen, H   carbon, C

FIGURE 1.4.12b: Different colours are used to represent different elements in circle diagrams.

Some atoms can chemically join together to form a **molecule**. Gases like oxygen exist as diatomic molecules, with the formula O_2. With the exception of the noble gases, all the elements that are gases at room temperature exist as diatomic molecules – for example hydrogen, H_2.

FIGURE 1.4.12c: Which molecules do these represent?

1. Why do we need to represent the particles of each element differently?

2. What is the difference between an atom and a molecule?

3. What is a 'diatomic molecule'?

Representing compounds

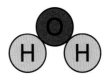

When atoms of different elements react, they chemically join together and form a new substance called a **compound**. The **ratio** of atoms in each unit of a compound is always the same and can be represented by a formula, for example H_2O.

> **4.** Represent the following compounds as circle diagrams:
>
> **a)** sodium chloride, NaCl   **b)** calcium oxide, CaO
>
> **c)** hydrochloric acid, HCl   **d)** calcium chloride, $CaCl_2$.

FIGURE 1.4.12d: The water molecule can be represented by a circle diagram, or by its chemical formula H_2O.

Thinking about atoms

The first person to use the idea of representing atoms as particles was Professor John Dalton. His atomic theory, presented in 1803, proposed a number of basic ideas:

- All matter is composed of atoms.
- Atoms cannot be made or destroyed.
- All atoms of the same element are identical.
- Different elements have different types of atoms.
- Chemical reactions occur when atoms are rearranged.
- Compounds are formed from atoms of the constituent elements.

Dalton gathered together all the ideas about chemical combination at the time, but he assumed that the simplest compound of two elements must be formed in a 1:1 ratio – for example water was HO. However, his work led other scientists to explore these ideas and provide new evidence to improve them.

FIGURE 1.4.12e: Dalton's atoms

> **5.** Describe what the circle models in Figure 1.4.12f show. (You are not expected to know the names of the atoms.)

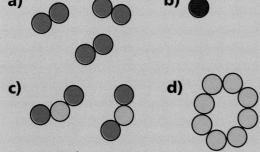

FIGURE 1.4.12f

> **6.** What does the formula of sulfuric acid, H_2SO_4, tell us?
>
> **7.** What assumptions did John Dalton make about atoms?

Key vocabulary

atom

element

molecule

compound

ratio

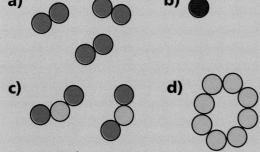

Understanding what happens when an element burns

- Make observations during chemical reactions.
- Write word equations to represent chemical changes.
- Explain chemical changes using a model.

Heat is a form of energy. By itself, heat can cause things to melt or boil, but not to burn. Burning requires oxygen as well as heat.

Physical and chemical changes

We heat gold to make it into new shapes. The gold melts; it does not react with oxygen and therefore does not burn. This is a physical change that is reversible. The liquid can cool, solidify and be melted again.

Burning is a chemical **reaction** with oxygen. When things burn, they are chemically changed into new substances. You cannot get back the original substance once you have burned it. The reaction is irreversible. Burning in air always produces compounds called oxides.

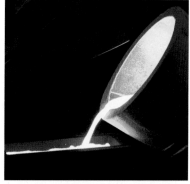

FIGURE 1.4.13a: Gold can be melted and solidified many times.

FIGURE 1.4.13b: When magnesium is heated in air it glows bright white and leaves a white powdery ash.

When magnesium is heated, we know that a chemical reaction has taken place because the appearance of the metal changes. The reaction also produces light energy and heat energy. This is a chemical reaction that can be represented as a word **equation**:

magnesium + oxygen ⟶ magnesium oxide

      REACTANTS   **PRODUCT**

The **reactants** are the substances that you start with and the new substances that are made are called the **products**. The products often look very different from the reactants.

1. What is the difference between a chemical change and a physical change?

2. How do you know that gold doesn't burn?

Rearranging atoms 》》

A chemical reaction involves **atoms** being rearranged and chemically joined to each other. We can represent this as a simple circle model or by using chemical symbols.

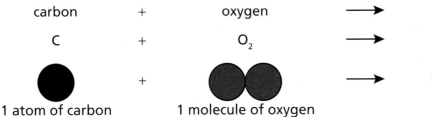

carbon	+	oxygen	⟶	carbon dioxide
C	+	O_2	⟶	CO_2

1 atom of carbon   1 molecule of oxygen   1 molecule of carbon dioxide

FIGURE 1.4.13c: Atoms are rearranged when a chemical reaction occurs.

3. Explain what is meant by 'burning'.

4. Write word equations for the burning of:

   a) zinc   **b)** sulfur.

Burning as a reaction 》》》

Two atoms of magnesium react with one molecule of oxygen to produce two units of magnesium oxide. The number of atoms of each element must be the same on both sides of an equation – this is called a balanced equation.

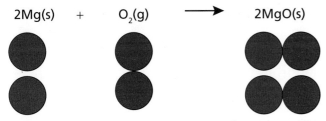

$$2Mg(s) \quad + \quad O_2(g) \quad \longrightarrow \quad 2MgO(s)$$

FIGURE 1.4.13d: Magnesium and oxygen react to form magnesium oxide.

5. What is meant by a 'balanced' equation?

6. Write a word equation and draw a circle picture for burning reactions that produce:

   a) potassium oxide, K_2O

   b) calcium oxide, CaO

   c) water, H_2O.

Did you know…?

The Hindenburg was a giant hydrogen-filled airship which travelled between Europe and America. In May 1937, just prior to landing, it burst into flames and was completely destroyed in less than a minute.

Key vocabulary

reaction

equation

reactant

product

atom

Observing how elements react in different ways

We are learning how to:

- Draw conclusions to explain observations.
- Use symbols and models to describe a chemical reaction.

How an element reacts and the products it makes are unique to the element and the conditions. These differences can help chemists to identify elements not only on Earth, but out in space, too.

FIGURE 1.4.14a: Fireworks use different metal compounds to produce different colours.

Using flames to recognise metals

Fireworks contain metal and non-metal compounds. It is the metals that are responsible for the colours – when they burn at high temperatures, they give off distinct colours.

TABLE 1.4.14: Flame colours of some metal compounds

Metal	Flame colour
potassium	lilac
sodium	yellow
calcium	orange/red
copper	green/blue
magnesium	white
strontium	red
barium	pale green

Carbon and sulfur combined act as a **fuel** that launches the firework and keeps it **burning**. When non-metals like these burn, they produce gases. Both sulfur dioxide and carbon dioxide dissolve in water to form **acids**. When metals burn in air the **products** are solid oxides that are the opposite of acids – these are called **bases**.

FIGURE 1.4.14b: Sodium burns with a yellow flame and produces a white powder called sodium oxide, Na_2O.

1. If an unknown compound turned a Bunsen flame green, what conclusion could you draw?

2. Which reactants will form the product sulfur dioxide?

3. What are the reactants if sodium oxide is produced?

4. How could you tell that it is sodium and not sulfur burning?

FIGURE 1.4.14c: Different oxide compounds

The temperature that substances are heated to is important. There has to be enough energy and air for elements to burn. The reaction not only forms an oxide, it also emits light. The colour of the light depends on the element, but the colour of the sparks produced by burning iron depends on the temperature. The colour changes from orange at lower temperatures to white when it is very hot.

Key changes indicate that a chemical reaction has happened:

- Has the substance permanently changed its appearance?
- Are there gas bubbles or a new smell?
- Has the temperature gone up or down? Can you feel heat?
- Change of pH – has an acid or a base been produced?

5. List the substances found used in fireworks and the reasons they are included.

6. List the evidence for burning sulfur being a chemical reaction.

7. Explain the difference between melting iron and burning iron in as much detail as you can.

8. Write a word equation for the burning of iron.

White light and smoke ▶▶▶

Zinc is used in fireworks to create smoke effects because zinc oxide (ZnO) is a non-toxic, fine white power. Zinc oxide is insoluble and is often used in sunscreens. Magnesium is used in fireworks and sparklers because it produces bright white sparks and ultraviolet (UV) light. UV light is emitted by the Sun and can damage your eyes. It is the reason you wear sunglasses and protect your skin with sunscreen.

9. What are the advantages and disadvantages of using magnesium in fireworks?

10. Use word equations and circle diagrams to represent the burning of zinc and magnesium in fireworks.

Did you know…?

Astronomers use special machines called spectrometers to study light that is emitted from distant stars and galaxies. Stars inside a nebula glow with beautiful reds, blues, and greens due to the different elements within the vast clouds of gas.

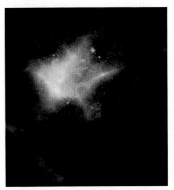

FIGURE 1.4.14d: Gas cloud nebula, galaxy and stars

Key vocabulary

fuel

burning

acid

product

base

Identifying the special features of carbon

We are learning how to:

- Explain the importance of carbon in our lives.
- Identify and explain the differences between an element and its compounds.

Carbon is an amazing element. It has the ability to form a huge number of different compounds with very different properties. Diamond and graphene are two of the hardest naturally occurring substances on Earth, while graphite is one of the softest. Our whole lives depend on carbon chemistry.

Carbonates and carbohydrates

Rocks and shells contain carbon compounds called **carbonates**. Calcium carbonate, $CaCO_3$, is found in chalk, limestone and marble and is used to make cement, concrete and glass.

Carbohydrates are found in plant materials and contain carbon, hydrogen and oxygen atoms. They are essential in our diet for energy and are found in foods like potatoes, rice, bread and pasta. Sugar or sucrose is also a carbohydrate and has the chemical formula $C_{12}H_{22}O_{11}$.

1. Which elements make up calcium carbonate and where is the compound found?

2. What is the difference between carbonates and carbohydrates?

3. What type of carbon compounds are:

    a) glucose ($C_6H_{12}O_6$)?   b) $CuCO_3$?

FIGURE 1.4.15a: Chalk is a carbon compound called calcium carbonate.

Organic compounds

Carbon atoms can form chains and rings – these compounds are called **organic compounds**. Methane (CH_4) is the simplest organic compound. Because it contains only hydrogen atoms and carbon atoms, it is called a *hydrocarbon* compound. It is a highly flammable gas that we use as a fuel in our homes and in the laboratory. Wood, petrol and heating oil all contain organic compounds with long carbon chains. When these fuels are burned carbon dioxide, water and soot are produced.

FIGURE 1.4.15b: These plastic items are made from polyethene, a hydrocarbon.

Polymers are used to make plastics. Polyethene can contain millions of carbon atoms and hydrogen atoms in very long chains. Polyethene is used to make packaging materials such as plastic bags and bottles.

4. What are hydrocarbons and what are they used for?

5. What is produced when hydrocarbons burn?

6. What is polyethene and what is it used for?

More materials made from carbon

Carbon has the highest melting point of all elements, 3 500 °C, and exists in two main forms. The graphite in your pencil has layers of atoms that can move, which makes it both soft and a very good electrical conductor. All the atoms in diamond are locked tightly into place, which makes it very strong. Diamond is so hard that it is used to cut and drill into other materials. Both these substances contain only carbon atoms, but the way in which the carbon atoms are joined together affects the properties of the material.

Carbon fibre is a flexible, light but very strong material that has many uses from crash helmets to mobile phone cases. Graphene consists of a single layer of carbon atoms arranged in a honeycomb structure and is almost transparent. It is 200 times stronger than structural steel, is a better thermal and electrical conductor than metals and is as flexible as plastic.

FIGURE 1.4.15c: Graphite and diamond are both carbon.

Did you know…?

Buckminsterfullerene is a tiny ball-shaped molecule containing 60 carbon atoms, C_{60}. These molecules are produced in lightning strikes and have been found in soot, rocks and in distant stars.

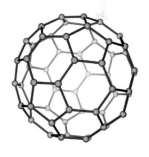

FIGURE 1.4.15e: Buckminsterfullerene molecule

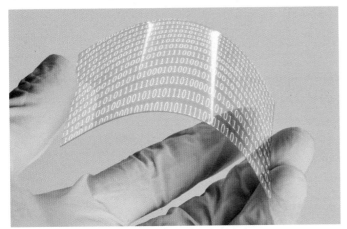

FIGURE 1.4.15d: Graphene is set to transform the electronics industry.

7. Explain the properties of diamond and graphite and why they are different.

8. Make a spider diagram to show the different uses of carbon and its compounds.

9. Why is carbon fibre a good replacement for metals?

Key vocabulary

carbonate

carbohydrate

organic compound

polymer

Understanding oxidation

We are learning how to:

- Describe oxidation.
- Recognise the effects of oxidation.
- Use data to support conclusions.

Oxidation is an important chemical reaction that causes big changes. Some oxidation reactions are fast and others are slow. Oxidation is sometimes useful and at other times it causes problems. Browning apples, rusting and burning are all oxidation reactions.

FIGURE 1.4.16a: The browning of an apple is an example of oxidation.

Oxidation

Oxidation is the name given to a chemical reaction in which oxygen is added to a substance. When a metal like copper is heated in air it reacts with oxygen. Black copper oxide is formed:

copper + oxygen ⟶ copper oxide

1. What is an oxidation reaction?

2. Give three examples of oxidation reactions.

3. What changes would you see as copper is heated in the Bunsen burner flame?

How much oxygen?

The chemical reaction between copper and oxygen can be used to measure the amount of oxygen in air. 100 cm³ of air is passed back and forth across some copper powder as it is heated strongly in a silica tube.

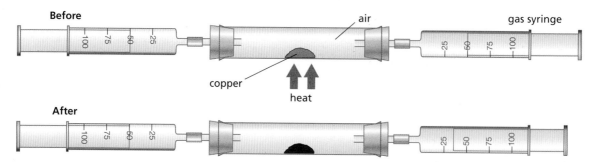

FIGURE 1.4.16b: Experiment to find out how much oxygen is present in air.

As the copper reacts with the oxygen, the volume of air decreases. When the volume stops changing the experiment is complete. The difference between the volume of air at the start and end of the experiment is the volume of oxygen. After the experiment, the mass of the powder in the tube is greater than the mass of copper at the start of the experiment.

4. Why does the volume of air decrease during the experiment?

5. Why does the volume of air eventually not decrease any further?

6. Describe and explain the change in appearance of the copper during the experiment.

7. Explain the change in mass of the powder in the tube.

Mass changes ▶▶▶

Some magnesium metal was weighed and then burned in a small heatproof container called a crucible. During the reaction the silvery metal changed into a white powder.

The results of the experiment are shown in Table 1.4.16.

FIGURE 1.4.16c: Heating magnesium, allowing air in but no oxide to escape

	At start	After heating	Change in mass
Mass of crucible and magnesium (g)	17.52	17.82	+0.30

TABLE 1.4.16: Experiment results

8. Explain the changes that have taken place during the experiment.

9. Write a word equation for the reaction.

10. Using circles to represent the atoms, draw a diagram to explain the change in mass.

Key vocabulary

oxidation

Investigating carbonates

We are learning how to:

- Describe the composition and uses of carbonate compounds.
- Recognise and explain thermal decomposition reactions.
- Identify carbon dioxide.

Calcium carbonate is one of the most versatile materials in the chemical industry. It is found naturally in chalk, marble, travertine and limestone. Calcium carbonate is the main component in egg shells, snail shells and the shells of shellfish like oysters.

Not just a rock

Calcium carbonate is added as a filler to paint, paper, plastics and toothpaste. It also helps relieve excess stomach acid and indigestion, and it is taken as a calcium supplement to maintain strong bones and teeth.

When heated, calcium carbonate forms lime (calcium oxide). This process is important in purifying iron. Lime is used to make cement and glass, and it is added to soils and water to reduce acidity.

1. Give two natural sources of calcium carbonate:

   a) a living source   **b)** a source from rock.

2. What is lime and how is it made?

3. Both calcium carbonate and lime are used to reduce acidity. Give an example for each substance.

FIGURE 1.4.17a: Calcium carbonate in the form of limestone is a common natural substance.

Heating carbonates

All **carbonate** compounds contain the same characteristic group of carbon and oxygen atoms.

calcium carbonate $CaCO_3$	magnesium carbonate $MgCO_3$	potassium carbonate K_2CO_3

These examples are all metal carbonates. Each unit of the compound contains a metallic element joined to the carbonate group. Some carbonates can withstand high temperatures because they are thermally **stable**. Most carbonates break down when they are heated. This kind of reaction is called **thermal decomposition**. 'Thermal' means heat and 'decomposition' means break down. Nothing is added in this type of reaction.

When carbonates decompose they produce carbon dioxide. For example:

copper carbonate $\longrightarrow$ copper oxide + carbon dioxide
(green solid)   (black solid)   (colourless gas)

$$CuCO_3 \text{ (s)} \longrightarrow CuO \text{ (s)} + CO_2 \text{ (g)}$$

As the carbonate is heated in the boiling tube, the carbon dioxide gas passes through the delivery tube into the **limewater** making it go cloudy.

4. What safety precautions should be taken during this experiment?

5. Describe the changes that you would expect to see during this experiment.

6. Why is this reaction called thermal decomposition? How do the equations show this?

7. What do the (s) and (g) symbols mean in the equation?

8. Before it was heated, the boiling tube and carbonate had a mass of 50 g. Explain why, after the experiment, the mass was less than 50 g.

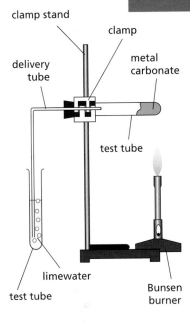

FIGURE 1.4.17b: This apparatus can be used to decompose a metal carbonate.

Making limewater ⟩⟩⟩⟩

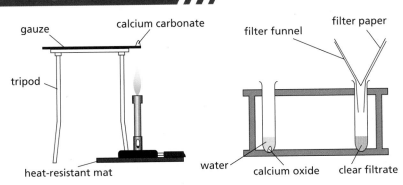

FIGURE 1.4.17c: If you heat a piece of limestone strongly it will decompose and form calcium oxide.

Some of the calcium oxide dissolves in water to form calcium hydroxide $Ca(OH)_2$ as shown in Figure 1.4.17c. This is called limewater.

If you add carbon dioxide to limewater, calcium carbonate is formed. Calcium carbonate is not very soluble so the limewater goes cloudy.

9. Write a word equation for the formation of limewater.

10. Draw a circle diagram for the thermal decomposition step and use it to explain the mass change during the process.

Did you know...?

Bicarbonate of soda contains sodium hydrogencarbonate, $NaHCO_3$. This is a raising agent, producing carbon dioxide in cake mixtures. The gas bubbles expand and escape as the cake cooks. This makes the light sponge texture.

Key vocabulary

carbonate

stable

thermal decomposition

limewater

Explaining changes

We are learning how to:

- Observe and explain mass changes.
- Use scientific terms and simple models to explain chemical processes.

Some changes in chemistry, such as melting and freezing, are reversible. These are are physical changes. Dissolving can usually be reversed by removing the solvent. Chemical reactions produce new products that are different from the substances you started with. Chemical changes are generally not reversible.

FIGURE 1.4.18a: Rusting is an oxidation reaction. It is a permanent change.

Chemical reactions

Oxidation and **thermal decomposition** are very different. Oxidation reactions involve oxygen being added to another substance. Compounds called oxides are formed:

metal + oxygen → metal oxide

Thermal decomposition reactions happen when substances like carbonates break down when they are heated:

metal carbonate → metal oxide + carbon dioxide

1. What is formed during an oxidation reaction?

2. Why is thermal decomposition a good name for this type of reaction?

Mass changes

During oxidation, oxygen is added. This means that the mass of the original substance will *increase*.

During thermal decomposition, carbon dioxide is produced. The gas escapes during the reaction and mixes with the air. The mass of the metal compound will *decrease*.

There are no mass changes when dissolving sugar because nothing is added and nothing escapes.

3. Explain why the mass changes for oxidation and thermal decomposition reactions are different.

4. Give two differences between dissolving and oxidation.

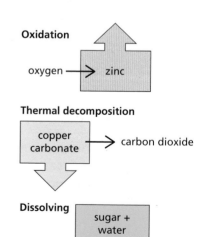

FIGURE 1.4.18b: Diagrams showing the mass change in different chemical processes

The Law of Conservation of Mass means that mass is never lost or gained in chemical reactions:

zinc	→	oxygen	+	zinc oxide
64 g	→	16 g	+	64 + 16 = 80 g

Reactants 80 g   Product 80 g

FIGURE 1.4.18c: The mass of the reactants is always the same as the mass of the products.

In the reaction in Figure 1.4.18c, the mass of the zinc increases as the two elements join to form zinc oxide. The total mass of the **reactants** is exactly the same as the mass of the new **product**. No additional mass is gained or lost – mass is conserved.

Some chemical reactions produce gases that escape and mix with the air. Even though no mass is lost or gained in the reaction, the mass can *seem* to decrease.

Heating carbonates or reacting them with an acid produces carbon dioxide. The mass decreases as the gas escapes. If you trap the gas in a balloon, the mass does not fall.

calcium carbonate	→	calcium oxide	+	carbon dioxide
$CaCO_3$	→	CaO	+	CO_2
100 g		56 g		44 g

FIGURE 1.4.18e

5. How does the circle model in Figure 1.4.18c help to show the conservation of mass?

6. Calculate the mass of the products in these reactions:

  a) magnesium + oxygen → magnesium oxide
      (24 g)   (16 g)

  b) S (32 g) + O_2 (32 g) → SO_2

7. a) Use the masses and the equations in Figure 1.4.18e to explain the conservation of mass for the thermal decomposition of calcium carbonate.

  b) If the gas was allowed to escape, what would the start and end masses be for this reaction?

  c) Explain how the symbol equation and the Law of Conservation of Mass are linked.

FIGURE 1.4.18d: No mass is gained or lost when a gas is produced and captured.

Did you know...?

The molecular mass of a compound is the sum of the atomic masses of all its component atoms. The formula tells you which atoms make up a compound. The atomic masses of all the elements are given in the Periodic Table.

Key vocabulary

oxidation

thermal decomposition

reactants

products

Checking your progress

To make good progress in understanding science you need to focus on these ideas and skills.

- [ ] Give some examples of elements, locate them in the Periodic Table and use the table to identify metals and non-metals.
- [ ] Give examples of elements and explain how they are organised in the Periodic Table.
- [ ] Define elements, use symbols, link the organisation of the Periodic Table to element features and explain how scientists organised the Periodic Table.

- [ ] Describe where some elements are found on Earth and identify some of the oldest known elements.
- [ ] Explain why different elements are found in different places and why they were discovered at different times.
- [ ] Use ideas and evidence to explain where and why elements and compounds were found.

- [ ] Identify some common properties of metal elements and non-metal elements and their uses.
- [ ] Classify metals and non-metals using their properties.
- [ ] Identify similarities and differences between metals and how these relate to their uses; compare and contrast properties of metals and non-metals.

- [ ] Identify metals and non-metals using data and suggest a reason for particular applications.
- [ ] Explain the properties of elements using data and why they are used for different applications.
- [ ] Select and justify the use of elements for different purposes, based on their properties.

- [ ] Describe an example of a compound and represent a chemical reaction using a simple model.
- [ ] Explain how compounds can be formed and explain a chemical reaction using simple models.
- [ ] Make links between simple models of compounds and chemical symbols.

- Identify changes during a reaction, relate these to reactants and products, and identify a difference between melting and burning.

- Make accurate observations, explain them using a simple model and a word equation and explain differences between chemical and physical changes in terms of atoms.

- Explain observations using word equations, relate chemical symbols to a simple circle model and use the correct terms and simple models to explain the differences between chemical and physical changes.

- Make observations and identify reactants and products.

- Make accurate observations, identify differences and, with support, describe reactions using simple models or word equations.

- Suggest reasons for different observations, describe reactions using word equations and start to use symbols to model chemical reactions.

- Recognise where carbon and its compounds are used.

- Explain different ways in which carbon is important.

- Explain, using the correct terms, where carbon is found and why it is useful.

- Identify oxidation and decomposition reactions.

- Explain why oxidation is a reaction; explain the differences between oxidation and thermal decomposition.

- Use models and word equations to explain changes during oxidation and thermal decomposition reactions.

Questions

Questions 1–7

See how well you have understood the ideas in the chapter.

1. What name is given to a reaction in which a chemical combines with oxygen? [1]

    a) aeration **b)** oxygenation **c)** oxidation **d)** breathing

2. What does the formula of carbon dioxide, CO_2, tell us about each molecule? [1]

    a) It has one atom of carbon and one of oxygen.

    b) It has two atoms of carbon and one of oxygen.

    c) It has one atom of carbon and two of oxygen.

    d) It has two atoms of carbon and two of oxygen.

3. What is a metalloid? [1]

    a) A cheap metal

    b) A recently discovered metal

    c) An element with some properties of a metal and some of a non-metal

    d) An alloy

4. If you were studying an element and wanted to find others with similar properties, should you look: [1]

    a) anywhere in the Periodic Table?

    b) up and down the groups?

    c) across the periods?

    d) somewhere else, because the Periodic Table won't help?

5. Why does a metal need to be extracted from an ore before it can be used? [2]

6. Give two examples of the differences between the properties of metals and non-metals. [2]

7. Explain, using examples, the differences between a chemical change and a physical change. [4]

Questions 8–14

See how well you can apply the ideas in this chapter to new situations.

8. Calcium carbonate can be used to relieve indigestion because: [1]

    a) it digests food   **b)** it tastes nice

    c) it increases acidity   **d)** it reduces acidity

9. What compound is formed when iron and oxygen react together? [1]

   a) oxyiron

   b) iron oxygen

   c) iron oxate

   d) iron oxide

a

b

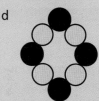
c

10. Which of the circle pictures in Figure 1.4.20a represents methane, CH_4? (Carbon is represented by the black circles and hydrogen by the grey.) [1]

11. Magnesium reacts with oxygen to form magnesium oxide. The mass of the magnesium oxide at the end of the experiment is greater than the mass of magnesium at the start because: [1]

   a) It burns with a bright light.

   b) It gives off carbon dioxide.

   c) Magnesium is not very dense.

   d) The oxygen has added to the mass of the magnesium.

d

FIGURE 1.4.20a

12. Draw diagrams to show the difference between a molecule of oxygen and a molecule of carbon dioxide. [2]

13. Stainless steel is an alloy of iron and chromium. It is used to make cutlery. Explain why it is a better choice of material than iron for this application. [2]

14. Which properties are important when making a bucket? In your answer suggest two possible materials, one metal and one non-metal, and compare them by referring to these properties. [4]

Questions 15–16

See how well you can understand and explain new ideas and evidence.

15. Figure 1.4.20b shows four substances: iron, sulfur, a mixture of iron and sulfur, and iron sulfide. Iron sulfide is formed by a fusion reaction that takes place when sulfur and iron are heated together. Draw and label a circle picture to represent each of the four substances. [2]

FIGURE 1.4.20b: Iron (top left), sulfur (top right), iron and sulfur mix (lower left) and iron sulfide (lower right)

16. Brass is a gold-coloured metal – it is an alloy of copper and zinc. It can be cast into different shapes and has a range of uses including musical instruments, electrical switches and door fittings. From this information suggest some of the properties of brass, explaining your answers. [4]

Forces and their Effects

Ideas you have met before

Gravity

Unless we support things, the force (pull) of gravity makes them fall to Earth.

Scientists such as Isaac Newton and Galileo Galilei helped us to understand gravity.

Contact and non-contact forces

Some forces need a contact between two objects, for example a hand pushing a door. Some forces, like gravity, do not need contact – they can act over a distance.

The pulling force from a magnet can attract certain materials from some way away.

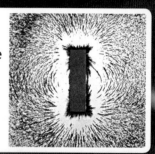

Friction

Friction is a force that acts between moving surfaces. When an object is moving, friction can cause it to slow down or even stop.

Air resistance is friction between air and an object moving through it. A parachute causes high air resistance and makes someone fall much more slowly.

A boat moving through water experiences water resistance. Boats are usually shaped so that they cut smoothly through the water.

Levers

Levers allow a small force to have a greater effect. For example, you could use a long metal tool like a screwdriver to lever a lid off a paint can. You probably could not produce a big enough force to do it with your bare hands.

Pulleys and gears also allow us to transfer forces in much more effective ways.

In this chapter you will find out

Types of force

Forces can stretch and compress things, or change the way something moves.

Forces can be turning forces as well as pushes or pulls.

We can draw force diagrams to help us understand the size and direction of forces and what effect they have.

We can use a newtonmeter to measure the size of a force.

Things that forces do

Elastic materials behave in a special way when a force changes their shape.

Materials can become permanently deformed when they are stretched or compressed by large forces.

A force on a moving object may cause its speed to increase or decrease.

The speed of an object depends on how far it travels in a certain time.

Useful and unwanted friction

Friction has many benefits and uses.

There are times when we want less friction.

Streamlining reduces frictional resistance.

Levers and turning forces

A lever works through a fulcrum to multiply a force.

By working out the size of turning forces we can make sure that structures balance.

Discovering forces

We are learning how to:

- Recognise different examples of forces.
- List the main types of force.
- Represent forces using arrows.

Forces are all around you, but you cannot see, touch or smell them. When forces cause movement you can see what they do, but when something is not moving there are still forces at work.

Types of force

A force can be a **pushing force**, a **pulling force** or a **turning force.** There is a pulling force from the Earth on this bungee jumper. Once he steps off the platform, the pulling force makes him fall. The arrow shows the pulling force making him move downwards. Without the pulling force of the Earth, he would not fall. The pulling force of the Earth on objects is called gravity.

FIGURE 1.5.2a: Gravity is a pulling force.

1. How would you describe the type of force that the Earth produces on the bungee jumper?

2. What is the name given to this force?

Multiple forces

A number of forces can be acting on something at the same time. The aeroplane in Figure 1.5.2b has four main forces acting on it:

- the downward pull of gravity

- the forward push from the engines

- the upward pull provided by the lift from the wings

- the pushing force of the air which resists the plane as it moves.

The direction of a force can be shown by an arrow. We can show how strong one force is compared to another by using different-sized arrows.

3. Which forces are helping the plane in Figure 1.5.2b to fly?

4. Which forces are working against the plane when it flies?

5. Which of the forces in the picture is the largest?

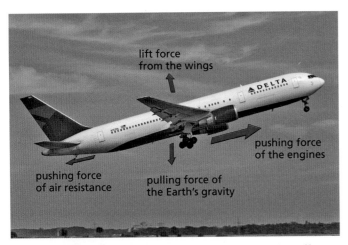

FIGURE 1.5.2b: These forces act on an aeroplane as it takes off.

Forces in balance

The two tug-of-war teams in Figure 1.5.2c are pulling equally and no one is moving. All the forces are in balance, which means each force is perfectly balanced by an equal force in the opposite direction. By accurately drawing the sizes and directions of the arrows on the diagram we can show that the forces are balanced. If the size of any one of the forces changes, the forces will no longer be in balance and there will be movement in the direction of the larger force.

FIGURE 1.5.2c: Forces are present, but there is no movement.

6. Explain what would happen to each of the forces if an extra person was added to one of the teams in Figure 1.5.2c.

7. Sketch a car that is starting to move away from a set of traffic lights. Draw arrows to show the forces at work.

8. Draw and explain the forces at work in these situations:

   a) a boat sailing across the sea

   b) a sledge being pulled over snowy ground.

Key vocabulary

pushing force

pulling force

turning force

Measuring forces

We are learning how to:

- Measure forces using newtonmeters.
- Use the correct unit for force.
- Explain the difference between mass and weight.

Forces are involved in some way in almost everything you do. Forces are acting on your body all the time. Some forces are huge, while others are tiny. What examples of really big forces can you think of?

Newtons and newtonmeters

Sir Isaac Newton is famous as an English scientist who lived in the 17th century. His ideas helped people's understanding of forces and are still very important today. In recognition of his work, the unit of measurement of force is called the **newton** (N). Instruments called **newtonmeters** are used to measure force.

FIGURE 1.5.3a: Sir Isaac Newton (1643–1727)

1. What is the unit of measurement for force?
2. What is the correct abbreviation for the unit of force?
3. What instrument is used to measure force?

Measuring with precision

Newtonmeters come in different models for measuring different-sized forces. In Figure 1.5.3b, one of the newtonmeters can measure force to a greater degree of **precision** than the other one, but it cannot measure such large forces. Selecting the appropriate measuring instrument is important for scientists. Smaller divisions on the scale allow more precise measurements, but instruments with small divisions can usually not measure large values.

FIGURE 1.5.3b: Two newtonmeters with different scales

4. What is the maximum value that each of the newtonmeters in Figure 1.5.3b can measure?

5. Which of the newtonmeters in Figure 1.5.3b allows the most precise reading? Explain your answer.

Weight, gravity and mass

The **weight** of an object is the force of **gravity** pulling down on the object. If there were no gravity then everything would be weightless. Because weight is a force, it should be measured in newtons. Weight can be measured using instruments such as newtonmeters and bathroom scales. Both give a reading of weight because the object being weighed is pulled down by gravity.

Mass is a measure of the amount of material in an object – the number of particles and type of particles it is composed of. Mass does not depend on the force of gravity, so it does not change if you take it somewhere where gravity is not as strong, such as the Moon. Mass is measured in kilograms. The mass of an object can be measured using a balance that compares the object with a known mass.

Sometimes people mix up 'mass' and 'weight', so scientists need to be careful to choose which term to use.

FIGURE 1.5.3c: Using a balance to find the mass of an object

> ### Did you know…?
>
> If you were in free fall you would feel as if you had no weight. In fact, if an object hanging on a newtonmeter were in free fall together, the weight reading would be zero. But mass does not change (except when you grow!).

6. Why do you think that some people confuse weight and mass?

7. If you measured the mass and the weight of an object on two planets of different sizes, what differences would you notice? Explain your answer.

8. Imagine a car crash on the Moon and the same crash on Earth. There would probably be no difference in the damage between the two crashes. Explain why this is the case.

Key vocabulary

newton

newtonmeter

precision

weight

gravity

mass

Understanding weight on other planets

We are learning how to:

- Explain the meaning of 'weightless'.
- Investigate weight on the Moon and on different planets.
- Identify the link between weight and gravitational attraction.

On planets with a smaller mass than Earth you would feel lighter and you could jump higher. On planets with a very large mass you would be so heavy that you wouldn't even be able to stand up. If you were in deep space and far from any planets or stars, you would have no weight at all.

Gravity in space

The force of **gravity** on you (your weight) depends on your distance from a planet. The further away you are from the Earth, the weaker the force pulling you back. In outer space, the distance to the nearest planets and stars could be so big that there would be no noticeable force of gravity and so you would be **weightless**.

1. In much of outer space there is little or no gravity. Why is this?

2. Think of a spacecraft setting off from Earth and travelling directly to the Moon. Describe the changes in gravity you expect the spacecraft to experience during the journey.

3. Suggest some differences you would experience when eating and drinking in weightless conditions, compared to on Earth.

FIGURE 1.5.4a: The Earth and the Moon. The bigger the mass of a planet or moon, the stronger its force of gravity.

Understanding gravity

Gravity is a force that pulls objects together. For example, your body is pulled towards the Earth, and the Earth and other planets are held in orbit around the Sun.

Gravity actually exists between *all* objects, but the force is only large enough to be noticeable when a massive object, such as a planet or a star, is involved.

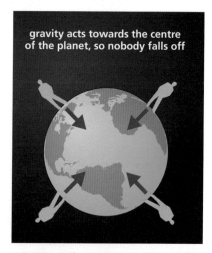

gravity acts towards the centre of the planet, so nobody falls off

FIGURE 1.5.4b: Gravity acts all over the Earth towards its centre.

TABLE 1.5.4: The effect of different values of gravity on the Moon and on other planets in the Solar System

	Earth	Moon	Mercury	Venus	Mars
Surface gravity (compared with the Earth's)	1.00	0.17	0.38	0.90	0.38
Your mass (compared with your mass on Earth)	1	1	1	1	1
How much you can lift (kg)	10	60	30	10	30
How high you can jump (cm)	20	120	53	22	53
How long it takes to fall back to the ground (s)	0.4	2.4	1.1	0.4	1.1

4. Look at Table 1.5.4. Where is gravity highest?

5. Using information from the table, write the planets and the Moon in order of increasing gravitational attraction if you were standing on the surface.

6. Explain why the mass of an object is the same on all planets and on the Moon.

A gravity puzzle

Gravity is an attractive (pulling) force between masses. What gravity would you experience if you tunnelled towards the centre of the Earth? Under the surface there would be a force of gravity from the mass of the Earth above you as well as from that below you. Because these forces are in opposite directions, the overall force of gravity would be lower than on the Earth's surface.

7. Imagine it was possible to build a tower on Earth to the height of an orbiting space station.

   a) What force(s) would you experience if you stepped off the tower?

   b) What movement would you expect?

8. Explain what would happen if you tried to weigh yourself in these situations:

   a) outer space

   b) in a tunnel, halfway to the Earth's centre

   c) on top of a tower at space station level

Did you know…?

When you see films of astronauts inside a space station orbiting the Earth, the astronauts appear to be weightless. But they, and the space station and everything in it, are actually still being attracted by the Earth's gravity. If there were no pulling force of gravity from the Earth, the space station would fly off into space.

Key vocabulary

gravity

weightless

Exploring the effects of forces

We are learning how to:

- Identify and describe the effects of forces of different sizes and directions.
- Predict and explain the changes caused by forces.
- Explain the concept of force pairs (action and reaction).

A spacecraft is hurtling through outer space. It is tempting to say that huge forces are involved, but in fact there are none.

A book is lying still on a table. People might think that there are no forces acting on the book, but there are.

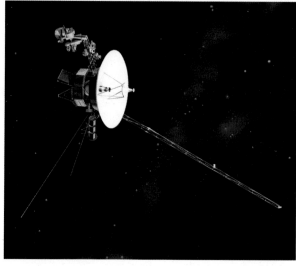

FIGURE 1.5.5a: What forces are acting on this spacecraft?

Forces causing movement 》

In Figure 1.5.5b, the trolley is still at first but will start to move as the mass is pulled downwards. When the mass reaches the floor, the string will go slack but the trolley will keep moving.

1. What force is acting on the hanging mass?

2. What forces are acting on the trolley:

   a) before it is moving?

   b) while the mass is dropping?

   c) once the mass has reached the floor?

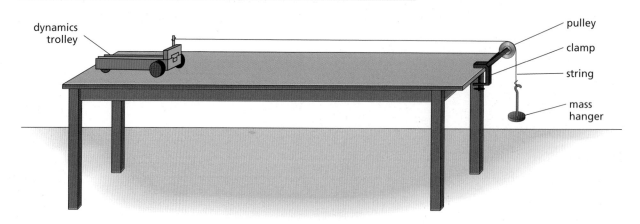

FIGURE 1.5.5b: Investigating the effect of force on motion

The effect of different sizes of force ⟩⟩⟩

If the trolley experiment in Figure 1.5.5b were repeated with a heavier hanging mass, a larger force would act on the trolley. The larger force would cause the trolley to move faster. In the same way, the harder a tennis player hits the ball, the larger the force on the ball and the faster the ball will leave the racket.

> **3. a)** Write a **prediction** about the movement of the trolley in Figure 1.5.5b, using this sentence stem: 'The heavier the mass hanging on the pulley, ...'
>
> **b)** Use the idea of force to explain your prediction.
>
> **4.** What do you think would happen to the speed of the trolley if, instead of increasing the mass hanging on the pulley, the mass of the trolley was increased?

Forces in pairs ⟩⟩⟩

When an object exerts a force on something else, it is called an **action force**. There is always an equal and opposite force called the **reaction force**.

Gravity is a pulling force between objects. A ball in the air is pulled towards the Earth, but the Earth is also pulled towards the ball!

When a person on a skateboard pushes against a wall, there is an action force against the wall, and the wall pushes back on the skateboarder with a reaction force. The two forces are equal in size and in the opposite direction to each other. When a cannonball is fired, the reaction force pushes the cannon backwards.

> **5.** In each of these situations, identify the action force and the reaction forces:
>
> **a)** a cricketer hitting a ball
>
> **b)** a person catching a ball
>
> **c)** someone opening a fridge door.
>
> **6.** Draw simple diagrams to show the forces when:
>
> **a)** a tennis player hits a ball
>
> **b)** a cyclist pushes down on a pedal
>
> **c)** a dog pulls on a lead.

FIGURE 1.5.5c: The reaction force from the wall pushes the skateboarder away.

Did you know…?

If an astronaut on a space walk released an object in space, it would float in the same position because there would be no forces on it. If the object was moving slightly when the astronaut let it go, it would keep going at the same speed and in the same direction.

Key vocabulary

prediction

action force

reaction force

Understanding stretch and compression

We are learning how to:

- Explain the relationship between an applied force and the change of shape of an object.
- Investigate forces involved in compressing and stretching materials.
- Identify applications for compressible and stretchable materials.

Imagine a mattress made of solid wood; clothes with no stretch; balls that don't bounce. The world would be a very different place if materials and objects could not change shape when a force is applied to them.

Comparing materials

All materials can **compress** (squash) or **stretch** to some extent. Some materials change shape by tiny, unnoticeable amounts – even with extremely large forces. Some materials may change shape with a small force but then break. When materials return to their original shape after the force is removed, this is called **elastic behaviour**.

1. Name some materials or objects that can be noticeably compressed or stretched *and* show elastic behaviour.

2. Name materials that show non-elastic behaviour when they are compressed or stretched.

FIGURE 1.5.6a: The angler benefits from elastic materials.

Size of force and amount of deformation

If you compress or stretch a material too far, it may not be able to return to its original shape – it remains deformed or it may break. In these situations the compressing or stretching force is beyond the **elastic limit** of the material.

Materials that break with a relatively small force (only slightly beyond their elastic limit) are said to be **brittle**.

3. Name some brittle materials.

4. Look at the data in Table 1.5.6. Write a list of features that a correctly plotted graph to show this data should include.

5. Plot a line graph to display the data in Table 1.5.6. Describe what your graph shows about how the force applied affects the spring.

TABLE 1.5.6: How a compressing force affects a spring

Force applied (N)	Compression of spring (cm)
0	0
10	3.1
20	6.2
30	9.3
40	12.4
50	15.5
60	16.1

6. a) From your graph, what do you notice about the compression when a force of 60 N is applied, compared to smaller forces?

b) What could explain the difference in part a)? Suggest why the final data point does not fit the pattern of the others.

Applications of elastic materials »»»

The elastic behaviour of springs make them ideal components in devices for measuring weight or force.

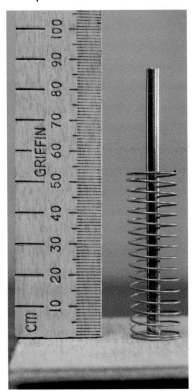

FIGURE 1.5.6b: The behaviour of a spring under an applied force allows us to measure weight.

Cushions on soft furniture, climbing ropes, clothing and the soles of sports shoes are all examples of uses of materials chosen for their elastic behaviour.

7. Why might a cushion not work well if the foam was:

a) too soft?

b) too hard?

8. Explain why springs are particularly suitable for use in weighing devices and forcemeters.

9. Suggest why a climbing rope would be less effective if it had no elasticity at all.

Did you know…?

The suspension in racing cars uses springs, which compress in a complex way depending on the force on them. This helps give the car good balance and grip at all speeds. The suspension also has dampers to help control the compression and bounce of the springs. Different springs and dampers are used at each race to suit the circuit and the track conditions.

Key vocabulary

compress

stretch

elastic behaviour

elastic limit

brittle

Investigating Hooke's Law

We are learning how to:

- Investigate the effects of applied forces on springs.
- Generate data to produce a graph and analyse outcomes.

Part of the skill of a scientist is collecting data and using it to find patterns that will improve our understanding of the world.

Springs are elastic

Elastic materials change shape when a force is applied to them, and then return to their original shape when the force is removed. The elastic behaviour of springs makes them useful in many situations – for example, a spring is used in a newtonmeter.

1. Give three uses of springs.

2. Suggest some properties of materials that would make good springs.

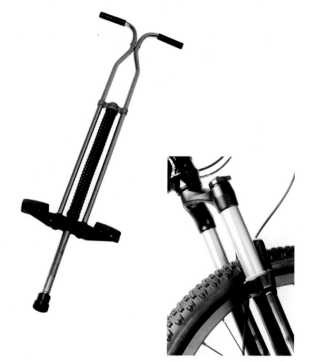

FIGURE 1.5.7a: Springs in action

Investigating how a force stretches a spring

Figure 1.5.7b shows the set up for investigating the stretching of a spring. As the force increases so does the spring's **extension**. 'Extension' means how much the spring has stretched, compared to its original length when no force was applied.

Within a certain range of forces, a spring will extend by regular amounts for equal increases in the force applied. So if a spring stretches by 1 cm when you apply a force of 1 N, then it will stretch by 2 cm if you apply a 2 N force. This behaviour of a spring is known as **Hooke's Law**.

3. A spring is being tested. It stretches by 3 cm when a force of 10 N is applied to it. If it behaves according to Hooke's Law, how far would you expect it to extend when these forces are put on it?

    a) 20 N   **b)** 70 N   **c)** 2 N

4. State Hooke's Law in your own words.

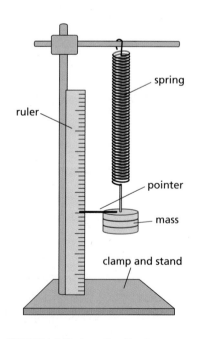

FIGURE 1.5.7b: Investigating Hooke's Law

Designing forcemeters ▶▶▶

A newtonmeter can only work accurately within a certain range of forces. This is because a spring stretches in even amounts, according to Hooke's Law, only up to a certain extension. Also, if even more force is added, the spring may not return to its original length when the force is removed. The spring has been stretched beyond its **elastic limit** and the device will be damaged and cannot be used again. Newtonmeters have an end-stop so that the spring cannot be stretched too far.

FIGURE 1.5.7c: This spring has been stretched beyond its elastic limit.

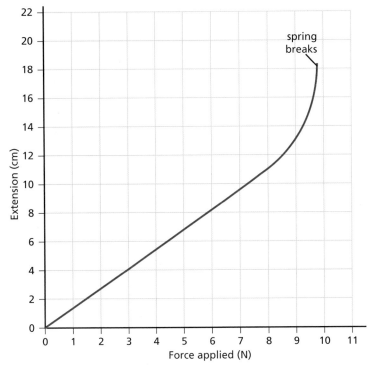

FIGURE 1.5.7d: The effect of a stretching force on a spring

Did you know…?

There are special phrases for describing relationships between variables in science and maths. When a graph, such as the one for Hooke's Law, has a straight line going through zero on both axes, we say that the two variables are *directly proportional*.

5. Look at Figure 1.5.7d. Describe what happens as the force on the spring is increased.

6. From the graph:

   a) how much force is needed to extend the spring by 7 cm?

   b) how much does the spring extend by if a force of 3.5 N is applied to it?

7. a) Suggest approximately what size of force is needed to exceed the elastic limit of the spring.

   b) Why is it not possible to be sure what the exact limit is from the graph?

Key vocabulary

extension

Hooke's Law

elastic limit

Understanding friction

We are learning how to:

- Identify the force of friction between two objects.
- Describe the effects of friction.
- Understand that friction acts in the opposite direction to the direction of movement.

Whenever there is movement, there is almost certainly friction. Friction is always working against movement.

Friction slows things down

The boy in Figure 1.5.8a is enjoying the thrill of the water slide. He picks up speed as he travels down so at the bottom he will make quite a big splash. The slide is designed so that **friction** doesn't slow people down too much.

- It's steep, so the boy's weight helps to overcome friction at the surface of the slide.

- The water pouring down the slide makes the smooth surface even more slippery.

1. What features of a water slide help people to travel fast? Explain your answer.

2. List three situations in which friction acts to slow something down.

FIGURE 1.5.8a: Is the friction on this slide large or small?

Friction is a force

Friction is a **contact force** that exists when two surfaces touch one another. Friction opposes movement. This means that the force of friction always acts in the opposite direction to movement, so it causes moving objects to slow down. When an object is stationary it will not move until the pushing, pulling or turning force is big enough to overcome the force of friction resisting movement.

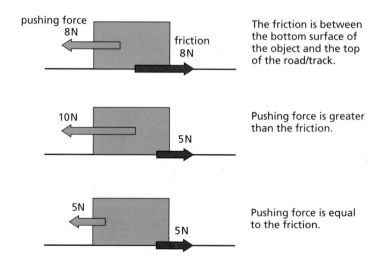

pushing force 8N
friction 8N
The friction is between the bottom surface of the object and the top of the road/track.

10N
5N
Pushing force is greater than the friction.

5N
5N
Pushing force is equal to the friction.

FIGURE 1.5.8b: Will the object move?

3. In general, in what direction does the force of friction act?

4. Look at Figure 1.5.8b. Assuming that in all three cases the object is stationary to start with, state in which case, if any, the object might move. Explain your reasoning.

Explaining friction

No surfaces are completely smooth, so when two surfaces touch each other the tiny bumps and ridges on one surface can rest in the hollows of the other. For the surfaces to slide across each other the bumps and ridges must ride up out of the hollows, and this needs force. The smoother the surfaces, the less force is needed to make the surfaces slide across each other.

Engineers take steps to reduce friction in machines. For example:

• they use a **lubricant** such as oil or wax to create a smooth sliding layer between moving parts

• they make surfaces as smooth as possible by special machining and polishing.

FIGURE 1.5.8c: The engine has to produce enough force to overcome friction.

5. Draw a diagram with labels to explain how friction occurs between two surfaces.

6. Draw an outline of a car in the middle of a sheet of paper. Around the outside add labels to identify where friction could occur. Add details to the diagram to make your descriptions clear.

7. Suggest some problems that friction could cause in a machine – for example a car or a bicycle.

Did you know…?

The oil in a car engine is very highly engineered to lubricate an engine, and so reduce friction between the moving parts. Engine oil has to work when the engine has just started on a freezing winter's day, and also at over 100 °C when the engine is hot. It has to withstand the constant pounding of all the moving parts in the engine.

Key vocabulary

friction

contact force

lubricant

Exploring the benefits of friction

We are learning how to:

- Describe applications that make use of friction.
- Design procedures for investigating the force of friction.

Friction, although frequently undesirable, can also have many benefits. In certain situations designers deliberately try to make friction high.

Situations in which sliding is bad >>

Without friction, many everyday activities would be very difficult.

- Your feet would slide backwards when you tried to walk.
- Cars could not accelerate, brake or go round corners because the tyres would slip.
- Clothes would slip through pegs on a washing line.

Friction resists movement when two surfaces might slide across each other, so it helps to provide grip.

1. Describe what might happen if you tried to open a door in a world without friction.

2. Explain whether friction is useful or unwanted in these situations:

   a) skiing

   b) rock climbing

   c) driving on a wet road

   d) pedalling a bicycle.

3. Draw force diagrams to illustrate the forces in action in the situations in question 2.

Increasing friction >>>

Buckles for webbing straps rely on friction of the strap against itself. The harder you pull on the strap the better it grips, because the two layers of webbing are forced together.

FIGURE 1.5.9a: How is friction helping in these examples?

In racing cars the brake discs are made of carbon ceramic. Friction causes so much heat during heavy braking that the discs glow red-hot. As the brakes get hotter, the friction of the carbon ceramic material increases. In wet weather the brake discs don't get hot enough to work well, so steel discs need to be used instead.

4. Explain how a webbing strap uses friction to stop itself coming undone.

5. Suggest why carbon ceramic disc brakes are more suitable for a racing car than for a family car.

Comparing grip >>>

The friction between an object and a ramp can be measured. If the friction is low, the object will start to slide when the ramp is raised slightly. If the friction is higher, the ramp must be raised higher before the object will start to slide. To compare different surfaces and get reliable results, a scientist must take care with the planning of the investigation.

FIGURE 1.5.9b: Why is the strap looped back through the buckle?

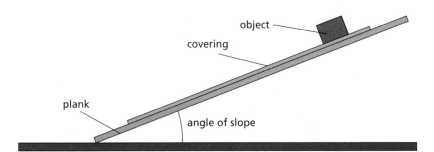

FIGURE 1.5.9c Measuring friction

6. Look at the apparatus for an investigation in Figure 1.5.9c.

   a) Identify the **independent variable**, the **dependent variable**, and any **control variables**.

   b) Explain how the experimenters could judge whether the results were **reliable**.

   c) State what **safety** considerations should be taken into account.

Did you know...?

When a car brakes, if the braking friction is too high it will overcome the friction that allows the tyres to grip the road. When this happens the car will skid. Once a car is skidding, it will stop quicker if the brakes are released briefly and re-applied. This allows the the tyre to regain grip on the road surface.

Key vocabulary

independent variable

dependent variable

control variables

reliable

safety

Understanding air and water resistance

We are learning how to:

- Link frictional forces between surfaces to 'drag' between objects in a fluid.
- Discuss examples of frictional drag in air and in water.
- Consider the effects of friction on sky divers.

If you open a car window when travelling at speed, you will notice how the air exerts large forces. Any object travelling fast has to overcome drag forces caused by the air. The faster you go, the greater force you must overcome.

Explaining air and water resistance

When you walk slowly, you may not notice the force that the air exerts on you. When you move faster you have to overcome more friction from the air. This force is also known as **air resistance** or 'drag'. When a car reaches its top speed, the air resistance is so large that it prevents the car going any faster. The pushing force from the engine is equal to the force of air resistance. The two forces are in balance so the speed cannot increase.

Water resistance works in a similar way. Water is denser than air so the drag is larger.

FIGURE 1.5.10a: The top speed of this car is limited by the power of the engine and the drag from the air.

1. What type of force is air resistance?

2. In what direction does air resistance act?

3. Why does air resistance limit the top speed of a vehicle?

4. What is the main difference between air resistance and water resistance?

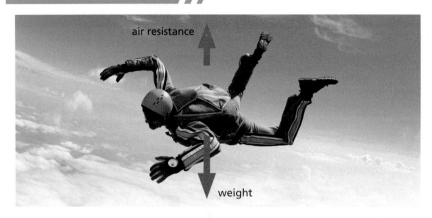

air resistance

weight

FIGURE 1.5.10b: The sky diver's speed initially increases because the weight force is larger than the force from air resistance.

When a sky diver steps out of a plane, he or she falls with increasing speed because of the force of gravity. As the sky diver falls faster, the air resistance increases. Eventually air resistance is so great that it balances the downward force of weight, and the sky diver reaches a steady speed known as **terminal velocity**.

When the sky diver opens the parachute, the air resistance increases greatly. The sky diver slows down dramatically and reaches a much slower steady speed.

5. At what point(s) during a sky dive are all the forces in balance?

6. When during a sky dive is the downward force of weight larger than the force of air resistance?

7. **a)** When during a sky dive is air resistance larger than the downward force of weight?

   b) Draw a force diagram to show the situation in part a).

Colliding particles ➤➤➤

Air is mixture of gases and consists of **particles**. When an object travels through air, it **collides** with the particles. These collisions make it more difficult for the object to move through the air. This is the cause of air resistance.

A fast-moving object collides with more particles than a slow-moving one, so the air resistance is larger at higher speeds.

8. Suggest why you cannot feel the collisions of the individual air particles on your skin.

9. Using the idea of particles, explain why it is much harder to run through water than through air.

Did you know…?

The fastest recorded sky dive was by Felix Baumgartner in 2012. He hit a peak speed of 1343 km/h after jumping from a helium balloon in the middle of the stratosphere. He reached such a high speed because the air is much less dense at the high altitude he jumped from, so there was little air resistance.

Key vocabulary

air resistance

water resistance

terminal velocity

particle

collide

Discovering streamlining

We are learning how to:

- Recognise natural and man-made examples of streamlining.
- Link streamlining to fuel efficiency in vehicles.
- Evaluate the use of data collected from investigations of drag.

Friction from air and water is a serious matter for anything that moves at speed – an animal, a plane, a car or a boat. Streamlining can increase speed and save fuel.

Streamlining in nature

Compare the shapes of the fish in Figure 1.5.11a. The shark is a predator that relies on speed to hunt in open water. The angel fish does not have the same need for speed because it feeds and hides in coral reefs. The shark's narrow, smooth shape, tapered at both ends, helps it to slip easily through the water with very little friction. We say that the shark's shape is **streamlined**.

If you compare the shapes of cars designed for high top speed with those that only travel slowly, you will find that fast cars are more streamlined. As well as increasing the top speed, streamlining means that less energy is needed to travel at a particular speed. Less energy is wasted.

FIGURE 1.5.11a: Compare the shapes of these two fish.

1. What features does a streamlined shape have compared to a less streamlined one?

2. State two benefits of streamlining.

3. Explain the effect of streamlining on friction.

Investigating streamlining

It is possible to compare the drag that happens with different shapes by timing how long it takes for objects to fall through a liquid. Using a thick liquid, such as wallpaper paste, means that the drag is high and the object will sink slowly. This makes it much easier to measure accurately than if we were dropping the objects through air or water.

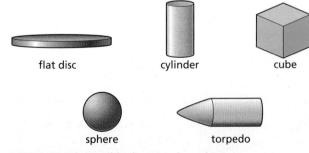

flat disc   cylinder   cube

sphere   torpedo

FIGURE 1.5.11b: Possible shapes to investigate

TABLE 1.5.11: Times taken for different shapes to sink

Shape	Trial 1: time taken to sink (s)	Trial 2: time taken to sink (s)
disc	19.4	25.1
cube	15.2	12.8
sphere	8.1	7.8
torpedo	8.9	9.6

4. Refer to the data in Table 1.5.11. Which shape is the most streamlined? Explain how you know.

5. **Evaluate** the data in Table 1.5.11.

    a) First suggest how reliable the results are.

    b) Now suggest factors that could have affected the results.

Using wind tunnels

Scientists and engineers use wind tunnels to investigate how the air flows over different shaped objects. Trails of smoke or tiny strips of paper show how the air is flowing. To get the best streamlining, the flow needs to be as smooth as possible. Where the flow has to change direction suddenly, **turbulence** is created and this increases drag. Drag can be reduced by having a smaller area meeting the air flow.

FIGURE 1.5.11c: Where is the flow streamlined and where is it turbulent?

6. Suggest two ways in which the choice of shape can reduce drag.

7. In designing a streamlined shape, why might the rear of the shape be important as well as the front?

8. What are the advantages to car designers of using wind tunnels to help design the shape of a car?

Did you know...?

Fluid dynamics is the study of how liquids and gases flow. This is very important in a wide range of situations:
- designing aircraft, boats and cars
- studying the oceans
- predicting weather patterns
- planning many industrial processes.

Key vocabulary

streamlined

evaluate

turbulence

Applying key ideas

You have now met a number of important ideas in this chapter. This activity gives an opportunity for you to apply them, just as scientists do. Read the text first, then have a go at the tasks. The first few are fairly easy – then they get a bit more challenging.

Adventure sport

The photographs show people doing the sport of paragliding. They launch from hillsides and glide through the air, like a paper aeroplane does. At its most basic, the sport involves flying from a high point downwards, at a constant speed, to a landing field a few kilometres away. More experienced paragliding pilots aim to fly as far as possible. They do this by finding rising currents of air, which they use to gain height. They can then glide further in search of more rising air. The world distance record stands at over 500 km for a single flight.

Figure 1.5.12a shows a person launching. They lay the paraglider on the ground and put on the harness. They then pull the glider up into the air and run down the slope. After a few steps, the glider has enough speed to fly and the pilot glides off the slope.

The harness is designed so that the pilot can sit comfortably and it also contains protective padding. Normally paraglider pilots land gently on their feet, but the padding helps to cushion them if they misjudge a landing.

Figure 1.5.12b shows a basic design of paraglider, as used by beginners. Figure 1.5.12.c shows a high-performance paraglider. The increase in performance is due to an improvement in the shape of the wing and a reduction of the drag caused by air resistance.

FIGURE 1.5.12.a: A paraglider launching

FIGURE 1.5.12b: A basic paraglider

FIGURE 1.5.12c: A high-performance paraglider

Task 1: Forces on a paraglider

Describe the forces acting on paraglider as it flies down at steady speed. What is important about the size of the forces if the speed and direction are not changing?

Task 2: Force diagram

Draw a force diagram to show the forces on a paraglider. Use the side-view diagram in Figure 1.5.12d as a basis. One of the force arrows has been drawn for you.

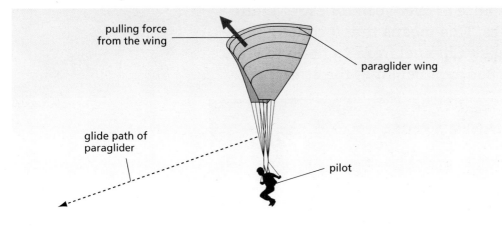

FIGURE 1.5.12d: Side view of a paraglider gliding

Task 3: Friction

Friction is essential to a paraglider pilot in many ways, but there is also unwanted friction. Explain how friction could be useful and where it is unwanted.

Task 4: Choosing the right elasticity

The designers must be careful to choose the right material for the protective padding in the harness. Thinking about how easy it is to compress the materials and how quickly it springs back – what might happen if the materials were not appropriate?

Task 5: Improving the performance of a paraglider

Suggest what features could improve the performance of a paraglider. Think about all the possible ways of reducing drag. There are other forces at work as well – could any of these help performance to be improved?

Task 6: Adding an engine

It is possible to have a harness with a small propeller engine on the back. This allows the pilot to gain height as they fly along. Explain how the forces on the wing might change and how this would affect flight.

Exploring forces and motion

We are learning how to:

- Recognise that for an object to start moving there must be a force applied.
- Describe the effects of balanced and unbalanced forces.
- Explain the significance of balanced and unbalanced forces on a moving object.

The glider in Figure 1.5.13a has no engine. It can glide a distance of over 50 m for every metre of height it loses. This means that it could glide across the English Channel from England to France with less than 800 m of height at the start.

FIGURE 1.5.13a: What forces are involved to allow this glider to glide long distances?

Forces in balance

Three forces are at work on a glider:

- a force from the wings which pulls on the glider in an upwards and forwards direction
- a downward force from the glider's weight
- a backward drag force caused by the glider's movement through the air.

Once the glider is flying, these three forces become **balanced**. The glider continues to glide on the same path at the same speed. Unless there is a change in any of the forces acting on the glider, its motion does not change. It will continue gliding until it reaches the ground.

1. What are the three forces that act on a glider as it flies?

2. What features does the glider in Figure 1.5.13a have that help it to fly as far as possible?

Did you know...?

Wings do not work in space because there is no air. With no air flow over them, the wings are incapable of producing any lift force. This is why spacecraft don't often have wings. One exception is the space shuttle – it needs wings to fly when it re-enters the Earth's atmosphere.

When forces are out of balance

If any of the forces on an object do not cancel out we say they are **unbalanced**.

- If the object is at first stationary, it will start to move.
- If the object is already moving, it will change its speed or its direction, or both.

3. Identify the similarities and differences between the forces acting in these situations:

  a) a toy car that is stationary on the floor

  b) a toy car rolling down a ramp at a steady speed.

4. Imagine someone pushing a toy car and letting it go so that it rolls across the floor. Describe how the forces change from start to finish.

A new balance

All the forces on the bridge in Figure 1.5.13b are in balance. The downward action force of the weight of the bridge is balanced by the upward **reaction force** of the ground, acting through the bridge supports. If a lorry parked on the bridge, the total downward force would increase. However, it would still be balanced, because the upward reaction force would now be larger. If the load was too great the bridge would collapse, or the compression force could break the supports.

FIGURE 1.5.13b: Engineers need to ensure all forces are in balance.

5. Sketch a road bridge, without any vehicles on it, and add arrows to show the forces acting.

6. Sketch your bridge again, now with a lorry parked on it. Add the force arrows for this new situation.

7. The bridge in Figure 1.5.13b spans a fast-flowing river. Describe or draw the additional force(s) that this creates.

Key vocabulary

balanced (forces)

unbalanced (forces)

reaction force

Exploring how forces affect speed and direction

We are learning how to:

- Recognise that the size of a force determines its effect.
- Recognise that the direction of a force determines its effect.
- Provide examples to illustrate where a force of precise strength and direction is needed.

If you could throw a ball on the Moon, it would follow a different path to the one it would follow on Earth. Doing the same in outer space would result in a different outcome again. Even if the ball was thrown with the same force, the forces acting on it in flight would be different in each case.

FIGURE 1.5.14a: An expert golfer can hit a ball into a small hole from many metres away.

Precise forces in action

In many situations, the speed and the direction of a force must be carefully controlled. A car driver controls the size of the pushing force from the engine using the accelerator pedal. The direction in which the force acts is controlled by the steering wheel. Many ball sports involve great control of the size of a force applied and its direction.

1. Why would a footballer want to use a different size of force in different situations?

2. What could go wrong if a person could not control the direction and size of the force when driving a car?

Hitting the target »»

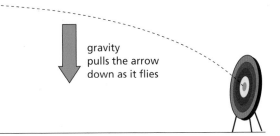

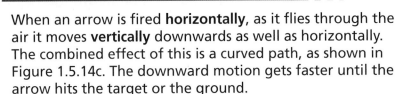

When an arrow is fired, the pushing force comes from the bow. Once the arrow has left the bow, two forces act on it as it flies through the air – the frictional force due to air resistance and the downward force of the arrow's weight. Air resistance is very small for an arrow because it is so thin and streamlined.

3. When an arrow is fired straight ahead (horizontally), what effect will the downward force of weight have?

4. What effect will air resistance have on the flight of the arrow?

Horizontal and vertical movement »»»

When an arrow is fired **horizontally**, as it flies through the air it moves **vertically** downwards as well as horizontally. The combined effect of this is a curved path, as shown in Figure 1.5.14c. The downward motion gets faster until the arrow hits the target or the ground.

gravity pulls the arrow down as it flies

FIGURE 1.5.14c: Something that moves through the air and is affected by the force of gravity is called a projectile.

Imagine the same arrow being fired on the Moon. Here the weight of the arrow is lower so although the arrow's path would still curve downwards, it would do so more gradually. Also, on the Moon there is no atmosphere to create air resistance, so the horizontal speed would stay the same as when it was fired.

5. Describe the motion of an arrow that is fired vertically upwards. Describe the force(s) that act after the arrow is fired.

6. Why does pulling the string on the bow affect how far the arrow flies?

7. What similarities and differences might there be between the flight of an arrow propelled by a bow and the flight of a football thrown by a person? Explain your answer.

FIGURE 1.5.14b: How does the archer know the direction in which to fire the arrow, so that it hits the target?

Did you know...?

In Olympic archery, the athletes fire at a target from a distance of 70 m. The bull's eye of the target is 12.2 cm in diameter, and they frequently hit this.

In 2010 a 14-year-old boy broke the record for firing an arrow the furthest distance. Zak Crawford's shot was nearly 500 m.

Key vocabulary

horizontally

vertically

Understanding speed calculations

We are learning how to:

- List the factors involved in defining speed.
- Explain a simple method to measure speed.
- Use the speed formula.

On Britain's busy roads, there are speed limits to make them safer. Driving too fast is one of the factors that causes accidents. Cameras that measure the speed of vehicles were introduced in the 1960s. In 2012 the number of deaths on Britain's roads was the lowest it had been since records began.

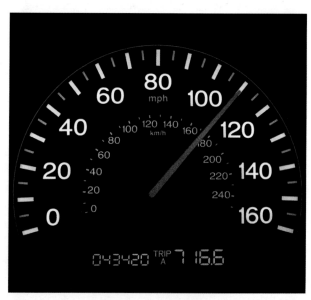

FIGURE 1.5.15a: This roadside camera measures the speed of a car.

Distance and speed

When you travel on a journey, it takes a certain amount of time to travel the **distance**. The **speed** of a vehicle is worked out from how how far a journey is and how long it takes. There are different **units** used for measuring speed:

- kilometres per hour (km/h)
- metres per second (m/s)
- miles per hour (mph).

When travelling fast your speed is high. You cover a longer distance in a certain time – you travel more kilometres in each hour, compared with travelling slower.

1. What does speed measure?

2. Which two quantities are needed to work out the speed at which something is travelling?

3. If car A travels 45 km in 1 hour and car B travels 55 km in 1 hour, which one has the higher speed?

4. Motorbikes C and D both travel 100 km. C takes 2 hours and D takes 3 hours. Which has the higher speed?

FIGURE 1.5.15b: A car's speedometer shows the car's speed at each instant.

We use a **formula** to calculate speed: | speed = distance travelled ÷ time taken |

The units of speed depend on which units were used for measuring the distance and the time.

Example calculation:
Usain Bolt from Jamaica won the 2012 London 100 metre final in a time of 9.63 seconds.

speed = distance travelled ÷ time taken

Usain's speed = 100 ÷ 9.63 = 10.38 m/s
This is equivalent to over 37 km/h or over 23 mph.

> **5.** Use the speed formula to calculate the speed of a cross-country runner who runs steadily for an hour and a half and covers 15 km. Show your working.
>
> **6.** A mouse runs 2 metres in 4 seconds. What is its speed?

Average speed »»»

> **Did you know…?**
>
> Some scientists have measured the force that an athlete's legs can produce, and how quickly the force can be transferred. From this they have worked out that it might be physically possible for the best athletes to run at over 60 km/h. We do not know if this will ever be achieved.

FIGURE 1.5.15c: For an Olympic sprinter the distance is measured in metres (m) and the time is measured in seconds (s), so the speed is calculated in metres per second (m/s).

When Usain Bolt won the Olympics sprint in 2012, his speed varied during the race. At the start it took a while to get up to full speed. The speed of 10.38 m/s that we calculated is his **average** speed over 100 metres. His top speed was over 12 m/s.

Some speed cameras work out a car's average speed over a distance of a kilometre or so, while other types work out speed almost in an instant. A car's speedometer displays the exact speed at any moment.

> **7.** Explain why your average speed and your top speed over a car journey will be different.
>
> **8.** What benefit to road safety may there be when cameras work out average speed over a distance, rather than in one spot?

Key vocabulary

distance

speed

unit

formula

average

Understanding turning forces

We are learning how to:

- Describe the forces acting on a see-saw.
- Understand that the forces turn about the fulcrum.
- Explain how to balance different weights on a see-saw.

When children play on a see-saw, they can balance each other by producing equal and opposite turning forces. Turning forces are involved in many situations – opening a door, riding a bike, driving a car, using a spanner.

Push and turn

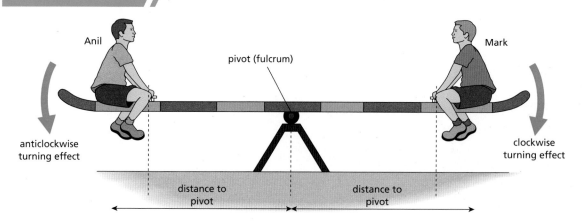

FIGURE 1.5.16a: Turning forces on a see-saw. Anil, on the left, causes an anticlockwise turning effect. Mark, on the right, causes a clockwise turning effect.

A see-saw changes a pushing force into a **turning force**. The force of the weight of a person sitting on the see-saw acts vertically downwards. This becomes a turning force because the middle of the see-saw sits on a **pivot**, also called a **fulcrum**.

For a door the fulcrum is the hinge. On a wheel the fulcrum is the axle at the centre of the wheel – a fulcrum does not change its position.

1. In what direction does the weight of a person act?
2. How does a see-saw change a vertical force into a turning force?
3. What acts as the fulcrum in a windmill?

The size of the turning force on a see-saw depends on the distance of the person from the fulcrum. A small pushing force can produce a large turning force when it is applied a long way from the fulcrum. When a large pushing force is applied close to a fulcrum it produces a relatively small turning force.

4. What would happen if the larger child on the balanced see-saw in Figure 1.5.16b moved closer to the fulcrum? Explain your answer.

5. If another two children, of equal weight, got on the see-saw in Figure 1.5.16b, in what positions could they sit and still get the see-saw to balance?

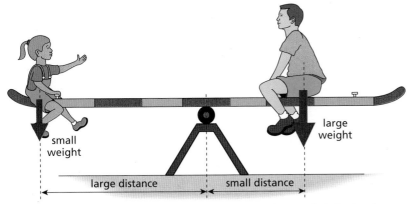

small weight

large weight

large distance   small distance

FIGURE 1.5.16b: The small force balances the large one because it is further from the fulcrum.

Force diagrams »»»

Straight arrows are always used to represent forces, even when they they result in a turning effect. The reason that a force acting in a straight line produces a turning motion is because of the fulcrum. The fulcrum, which is some distance from where the force is acting, acts as a fixed point. The turning motion occurs around the fixed point.

6. Draw and label forces diagrams for:

   a) a person pushing open a door

   b) a spring 'door-closer' closing a door

   c) a person forcing a door open against a spring-closer.

7. Explain why a person pushing with a smaller force than the spring can still open the door in question 6.

8. Give some examples of turning forces in action in a science laboratory.

Did you know...?

By positioning the forces and the fulcrum carefully, jacks allow people to raise a heavy car by using just one hand to apply a force. This works because where the person pushes is a long distance, compared to the distance of the car, from the fulcrum.

Key vocabulary

turning force

pivot

fulcrum

Discovering moments

We are learning how to:

- State and use the law of moments.
- Describe how turning forces can be increased.
- List some examples of levers used as force multipliers.

A turning force is known in physics as a 'moment'. By understanding moments we can use levers to produce large turning effects.

Working out a moment

The size of a **moment** depends on the the size of the force used and on the distance of the force from the fulcrum. The equation for working out the size of a moment is:

moment = size of force × distance of the force from fulcrum

The unit is newton metres (Nm).

On a see-saw there are two moments at work. One of them turns the see-saw clockwise and the other turns it anticlockwise. For a balanced see-saw:

clockwise moment = anticlockwise moment

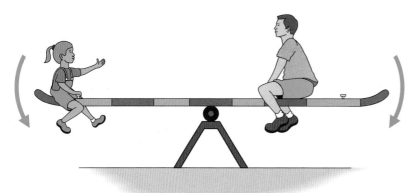

FIGURE 1.5.17a: When the clockwise and anticlockwise moments have the same value, the see-saw balances.

1. What does the scientific term 'moment' mean?

2. What two quantities do we need to know to be able to work out the size of a moment?

3. Look at Figure 1.5.17b. Explain why it is easier to undo a nut with a long spanner than with a short one.

4. A force of 20 N is pushing on a see-saw 2 m from the fulcrum. Calculate the size of the moment.

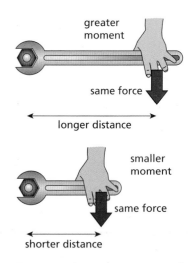

greater moment

same force

longer distance

smaller moment

same force

shorter distance

FIGURE 1.5.17b: How a spanner works

Levers

A **lever** is an example of a very simple machine. Levers use moments to produce large forces. This means that a large **load** can be moved by a smaller **effort**.

In Figure 1.5.17c the person's downward push gives an anticlockwise moment on the left of the fulcrum and this balances a clockwise moment on the righthand side.

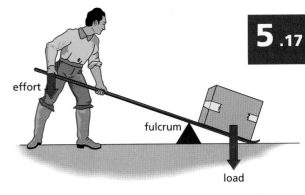

FIGURE 1.5.17c: A see-saw being used as a lever

5. In Figure 1.5.17c, the load is 1 m from the fulcrum and gives a downwards force of 20 N.

   a) What is the size of the anticlockwise moment caused by the load?

   b) What moment must the person produce to overcome the moment of the load and lift it?

   c) What force would the person need to push down with if they were 2 m from the fulcrum?

Simple machines

Levers that work like the one in Figure 1.5.17c act as **force multipliers**. These simple machines make a job easier. Examples include: a crowbar, a nutcracker, a spanner, long-handle garden cutters, a wheel barrow.

FIGURE 1.5.17d: Examples of levers used as force multipliers

6. Explain why a lever can be called a 'force multiplier'.

7. Explain how garden cutters work as a lever. Use the key words and a force diagram in your explanation.

8. A nutcracker works slightly differently. Try to draw a force diagram for one. What is the difference between this and the other levers?

Key vocabulary

moment

lever

load

effort

force multiplier

Understanding the application of moments

We are learning how to:

- Link the law of moments to the design of cranes.
- Explain why counterweights are needed by cranes.
- Investigate the lifting capacity of a crane.

Engineers use their understanding of moments to design many types of machine – cranes are one example of this. Cranes make many tasks easier in building work. Ancient structures such as the Egyptian pyramids and Stonehenge are all the more remarkable for the fact that they were built without the use of such modern machinery.

FIGURE 1.5.18a: The pyramids were built without using cranes.

Keeping cranes stable

The larger the load that a crane lifts, the more it is in danger of toppling over. Also, the further crane has to reach to lift a load, the greater the risk of toppling. This is because of the size of the moment caused by the load. The tower crane in Figure 1.5.18b has a **counterweight** on the non-lifting end, to balance the load. This allows the crane to lift much bigger loads safely.

1. Why are tower cranes at risk of toppling over?

2. Describe one feature of a tower crane that helps in stopping it from toppling over.

3. Why is the size of the counterweight important?

Adjustment for different loads

A crane is often required to lift different-sized loads. This could unbalance the crane, because the size of the moment changes. By moving the counterweight to a new position along the jib, the clockwise and anticlockwise moments can be balanced.

A crane may sometimes have to reach further out than at other times. Again in this situation the position of the counterweight needs to be adjusted to keep the crane balanced.

4. Explain why a moveable counterweight is necessary on a crane.

5. A crane has just lifted a small load. In what direction does the counterweight need to be moved if the crane now has to lift a larger load?

6. A crane has picked up a load from the far limit of its reach and now needs to release it near to the base of the crane. How can the crane be kept balanced?

Exact position of the counterweight ⟫⟫

For all the different lifting work that a tower crane does, it is important that the counterweight is always moved to the correct position along the jib. By calculating the clockwise and anticlockwose moments, engineers can work out the exact position for the counterweight.

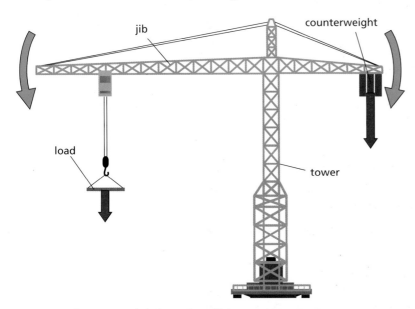

FIGURE 1.5.18b: Forces and their turning effects on a tower crane

7. What causes the anticlockwise moment on the crane in Figure 1.5.18b?

8. If the counterweight weighs 1000 N and each side of the jib is 20 m long, calculate:

    a) the maximum possible clockwise moment

    b) how big a load could be lifted when the counterweight is 5 m from the tower and the load is 10 m from the tower.

9. If the load and the counterweight were kept perfectly balanced, what other factors could affect the maximum lifting capacity of a crane?

Did you know…?

A crane in a shipyard in China can lift over 20 000 metric tonnes, which is 20 000 000 kg. This is roughly the weight of 15 000 medium-sized family cars.

Key vocabulary

counterweight

Checking your progress

To make good progress in understanding science you need to focus on these ideas and skills.

- List types of force and represent forces using force diagrams; use newtonmeters.

- Describe the size and direction of forces using force diagrams.

- Explain how the size and direction of forces determines their effects.

- Identify gravity as a pulling force and distinguish between mass and weight.

- Describe what is meant by mass, explain how gravity forces affect weight, explain why weight varies from planet to planet and explain the term 'weightless'.

- Explain weight as a gravitational attraction between masses which decreases with distance; use scientific concepts to explain the difference between mass and weight.

- Know that forces can lead to changes in shape and investigate the change of shape of a spring.

- Explain the relationship between the amount of change in shape and the size of the force, and use data to state Hooke's Law.

- Collect accurate data about forces changing the shape of an object, recognise when shape changes regularly with force size, and explain behaviour when the elastic limit is exceeded.

- Identify some situations in which forces are balanced and recognise that unbalanced forces are needed for a change to take place.

- Identify forces acting in pairs, and apply understanding of forces to explain how a force can cause a change in speed and direction.

- Identify different examples of forces and reaction forces, and predict the changes of speed and direction that different forces can cause.

Recognise that friction is a force that slows objects down or stops them from moving.

Explain that friction is a contact force opposing the direction of movement.

Provide a detailed explanation of friction between surfaces.

List examples in which friction is useful and when it is unwanted, recognise that drag forces slow things down, and recognise that streamlining helps objects move through air or water.

Compare contrasting situations involving friction, explain how friction can be increased or reduced, explain air and water resistance, and explain how streamlining reduces such resistance.

Explain air and water resistance in terms of frictional drag, explain the forces on flying or falling objects, and explain streamlining using scientific vocabulary.

Explain how to find the speed of an object.

Explain the concept of speed and use an understanding of speed to explain how the equation for speed is derived.

Independently derive the equation for speed and use understanding of the speed equation to explain how speed cameras work.

Describe the balancing of a see-saw with different loads, recognise situations in which balance is important, and describe the effect of increasing the length of a lever.

Explain how a fulcrum allows a turning motion, explain the effect of changing the size of a force or its distance from the fulcrum, and use and apply the law of moments.

Explain moments using force diagrams and the law of moments, explain how levers can act as force multipliers, and explain and demonstrate the design principles of a crane.

Questions

Questions 1–7

See how well you have understood the ideas in this chapter.

1. Which is the best description of gravity? [1]
    a) It pushes you towards the Earth.
    b) It is a pulling force between two objects.
    c) It is force that is stronger if you are on a planet with less mass.
    d) It is a pushing force between two objects.

2. What is the speed of a cyclist who covers 7 m in 1 second? [1]
    a) 70 km/h   **b)** 7 km/h   **c)** 7 m/s   **d)** 700 m/s

3. What is the correct unit for measuring force? [1]
    a) metre   **b)** newton metre   **c)** newton   **d)** centimetre

4. Which of these statements about forces and movement is true? [1]
    a) If an object is moving there must be a force acting.
    b) When a parachutist opens his parachute he moves upwards.
    c) Balanced forces cause movement.
    d) A moving object continues at the same speed unless an unbalanced force acts.

5. Describe the forces acting on a trolley as it rolls down a ramp. [2]

6. Describe why streamlining is important for a car. [2]

7. Explain the forces involved when two people of different weights balance each other on a see-saw. [4]

Questions 8–14

See how well you can apply the ideas in this chapter to new situations.

8. Two people are pushing either side of a door. One is pushing with a force of 80 N, 100 cm from the hinge. The other is pushing 50 cm from the hinge. Which size of force must this second person push with to balance the other? [1]
    a) 160 N   **b)** 40 N   **c)** 800 N   **d)** 8 000 N

9. Which of these statements is true about your weight and mass on a planet that has twice the gravity of Earth? [1]
    a) Weight is the same, mass is double.   **b)** Weight and mass are both the same.
    c) Weight and mass are both double.   **d)** Weight is double, mass is the same.

10. A speedboat takes 8 seconds to pass from one buoy to another 56 m away. Which of these is the speed of the boat? [1]

    a) 448 m/s   **b)** 7 km/h   **c)** 7 m/s   **d)** 448 km/h

11. For a large object dropped onto a trampoline, which of these statements is *not* true? [1]

    a) The trampoline shows elastic behaviour.
    b) The object's weight exerts a downward force on the trampoline.
    c) The trampoline exerts an upward force on the object.
    d) The trampoline exerts a downward force on the object.

12. Draw and label a diagram to show the forces acting on an object that has just been dropped from a hot air balloon. [2]

13. Draw a diagram to show the forces acting on an object falling towards Earth at terminal velocity. Explain why the object is at terminal velocity. [2]

14. The soles of running shoes are made from an elastic material. Explain what is meant by 'elastic' and how this helps the runner. [4]

Questions 15–16

See how well you can understand and explain new ideas and evidence.

15. Evaluate the following statement, stating the extent to which you agree with it: 'A jack designed to lift a car could be strong enough to lift a lorry'. [2]

16. Crash-helmet padding is compressed as a force is applied. This protects the head because the squashing reduces the effect of an impact. Figure 1.5.20 shows how three different types of padding for motorcycle crash helmets gets compressed when different forces are applied. Evaluate the three types, suggesting which is most suitable. [4]

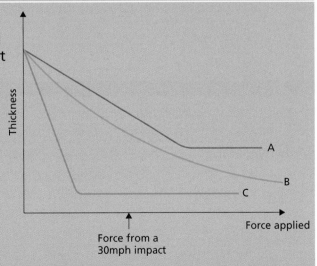

FIGURE 1.5.20: A graph to show the effect of thickness of crash-helmet padding when different forces are applied.

Energy Transfers and Sound

Ideas you have met before

Gravity

Gravity is a force that pulls objects towards the Earth. When they fall, they speed up – think of a bungee jumper, freewheeling on a bike down a hill or riding on a roller coaster.

Sound and vibrations

Sounds are only possible when a vibration occurs. Banging on a drum or plucking a guitar produces vibrations that cause a sound to be made.

Sound and volume

We can change the vibrations of a sound by giving them more energy. The stronger the vibrations, the louder the sound. When you pluck a guitar string lightly, a soft sound is made. Pluck it very hard and a much louder sound can be heard.

Sound and pitch

Some sounds we hear have a high pitch, like a whistle or a siren. Some have a low pitch, like the rumble of thunder. When we change the pitch, we change how rapidly an object vibrates.

Energy

- Energy makes things happen.
- Energy can be stored in something that is high up, as gravitational potential energy. This is transferred as movement energy when the object moves downwards.
- Energy can be transferred in many different ways. Useful energy transfers make life easier.

Useful and useless energy transfers

- Energy can be transferred by different processes. Some transfers result in useful changes, such as using an energy-efficient light bulb.
- In other situations useless energy transfers occur, for example the noise from a hairdryer.

Transferring more energy

- Fuels are special chemicals that release a lot of heat energy by burning.
- Different fuels store and transfer different amounts of energy.

Energy is carried by sound

- Sound energy is transmitted by waves (vibrations) being passed on by air particles.
- Echoes occur when sound waves are reflected by hard materials.
- The ear is designed to capture sound waves.
- Many animals communicate with sounds that we cannot hear. We have found useful applications for 'ultrasound' and 'infrasound'.

Exploring energy transfers

We are learning how to:

- Recognise what energy is and its unit.
- Describe a range of energy transfers using simple diagrams.
- Use a Sankey diagram as a model to represent simple energy changes.

The Sun is our main source of energy. Plants convert this energy by chemical processes to make food. Solar panels transfer the Sun's energy by electric current to provide electricity for our use. By transferring energy from the Sun, useful energy can be provided for our planet.

FIGURE 1.6.2a: Where has the energy to light this bulb come from?

What is energy?

When energy is transferred, useful things can happen. When a log is burned, energy is transferred by chemical reactions to the surroundings by light and heat. Switching on a light bulb transfers energy by electric current to the bulb. Energy is then transferred from the bulb to the surroundings by light and heat.

Energy is never lost or made, it is just transferred by different processes to different places. We measure it in a unit called a **joule** (J).

1. Look at the photos on this page. In which of them is energy being transferred?

2. **a)** What is happening as a result of the energy transfer you can see in Figure 1.6.2b?

   b) What is happening in the other photo in Figure 1.6.2b? Why is it not possible for energy to be transferred here?

Energy transfers

It is useful to track the processes by which the energy is transferred. This can be done using a simple **energy transfer diagram** (see Figure 1.6.2c). When you switch on a light bulb, you want to transfer energy by light. However, the light bulb also gets hot. Transferring energy by heat is not a useful change in this instance. Energy-efficient light bulbs have been designed to transfer more energy by light and less by heat.

FIGURE 1.6.2b: Describe the differences, in terms of energy transfer.

FIGURE 1.6.2c: Simple energy transfer diagram for a light bulb

3. Write a sentence to describe the energy transfers shown in Figure 1.6.2c.

4. Draw a diagram to show how energy is transferred by:

   a) a boiling kettle   b) a toaster   c) a log fire

5. In your answers to question 4, underline the useful energy transfers and circle the unwanted energy transfers.

Did you know...?

The amount of energy transferred to the Earth from the Sun in one minute is enough to meet the world's energy demands for one year.

Sankey diagrams

If you move a weight of 1 N through a distance of 1 m, you transfer 1 joule (1 J) of energy. One joule of energy is also needed to heat 1 cm³ of water by 1 °C.

A **Sankey diagram** is an energy transfer diagram that shows the relative amounts of energy transferred by a device. The width of each arrow shows how much energy is transferred. The non-useful energy transferred is always shown pointing downwards.

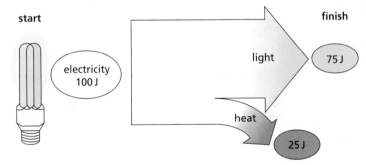

FIGURE 1.6.2d: Sankey diagram for an energy-efficient light bulb. How would the Sankey diagram for an old-style, less efficient light bulb compare with this one?

For example, in Figure 1.6.2d, 100 J of energy is transferred to the light bulb by electric current. It transfers 75 J by light (useful) and 25 J by heat (non-useful) to the surroundings. If you draw these on graph paper, you can accurately represent the proportions of energy involved.

6. On graph paper, draw a Sankey diagram for an electric drill that transfers 500 J of electricity by 100 J of sound, 100 J of heat and 300 J of movement energy. You will need to decide which of the outputs are useful and which are useless.

7. How could you make the drill in question 6 more efficient?

Key vocabulary

joule

energy transfer diagram

Sankey diagram

Understanding potential energy and kinetic energy

We are learning how to:

- Recognise energy transfers due to falling objects.
- Describe factors affecting energy transfers related to falling objects.
- Explain how energy is conserved when objects fall.

Many theme parks make use of energy transfer in their rides. An object high up has the potential to transfer energy. There are plans for a new vertical-drop ride, the 'Drop of Doom', which, at 126 metres tall, will be the tallest ever. People will fall from a stationary position at the top and reach speeds of up to 200 km per hour.

What is gravitational potential energy? »

Objects at a height possess energy, because of **gravity** – think of parachute jumpers or sky divers. This energy is known as **gravitational potential energy**.

You use this energy when you cycle down a hill, ride a zip wire or go on a roller coaster.

1. What is the unit for gravitational potential energy?
2. What is the name of the force acting on objects that causes them to have gravitational potential energy?

Factors affecting gravitational potential energy »

The higher an object is the more gravitational potential energy it has. More energy can be transferred to make it move. When the object falls, energy is transferred by **kinetic energy**. The object moves faster as more energy is transferred.

FIGURE 1.6.3a: How is energy being transferred as people drop from the top to the bottom of this ride?

FIGURE 1.6.3b: How does gravitational potential energy affect these people?

The greater the force acting on the object, the more energy can be transferred. The force of gravity is greater on Jupiter than on Earth, so an object falling the same distance on Jupiter will transfer more gravitational potential energy than it would on Earth.

3. A tennis ball falls from the following heights. Which will transfer the most gravitational potential energy?

   a) 10 mm   b) 10 cm   c) 10 m

4. Look at Table 1.6.3. If a ball of mass 2 kg is dropped from a height of 1 m on each planet, on which planet will it reach the highest speed?

TABLE 1.6.3: Gravitational strengths on different planets

Planet	Gravitational strength (N/kg)
Earth	9.8
Mercury	3.6
Mars	3.7
Saturn	11.3

Conservation of energy in falling objects »»»

Gravitational potential energy is transferred by movement and heat. As a falling object drops lower, its gravitational potential energy decreases and the amount of energy transferred to kinetic energy increases. Some energy will also be transferred to the surroundings by heat, due to friction with the air particles during the fall. The faster the object falls, the greater the energy transferred by heat. When the object hits the ground, all the kinetic energy is transferred by heat and sound to the surroundings.

5. Look at Figure 1.6.3c of a ball falling from a height. In which position (A, B or C) does the ball have:

   a) the highest gravitational potential energy?

   b) the lowest gravitational potential energy?

   c) the lowest kinetic energy?

   d) the highest kinetic energy?

6. Sketch two graphs to show how the gravitational potential energy and the kinetic energy of the ball in Figure 1.6.3c change during the fall.

Did you know...?

The Three Gorges Dam in China is the biggest use of gravitational potential energy in the world. It is a hydroelectric dam that uses water stored at about 175 m above sea level. Its gravitational potential energy is transferred to produce about ten per cent of China's electricity output.

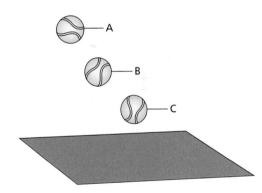

FIGURE 1.6.3c: A ball transferring gravitational potential energy

Key vocabulary

gravity

gravitational potential energy

kinetic energy

Doing work

We are learning how to:

- Recognise situations where work is done.
- Describe the relationship work done = force × distance.
- Apply the equation for work done to different situations.

Ancient Egyptians used a 'shadouf' to lift heavy buckets of water from deep rivers. A shadouf is a type of lever – a simple machine that uses a force to transfer energy. Machines help us to do work.

FIGURE 1.6.4a: A shadouf, lifting water from a lake

Linking energy and force

A force can transfer energy. If you were pulling a heavy load along, you would need to use a large force. Energy is transferred in this situation. Chemical energy from your muscles is transferred by kinetic energy and heat. We say that work is done when a force is used to transfer energy.

The **work done** is equal to the energy transferred, and is measured in joules (J).

1. Which situation in the photos needs the biggest force?

2. Draw an energy transfer diagram for both situations in Figure 1.6.4b.

Defining work done

The further you pull or push a load, the more work you do. If two people were pushing identical boxes along a floor but one person pulls it twice the distance, that person will do twice as much work. Or if one box is much heavier and needs double the force to push, then double the amount of work will be done. The work done depends on the size of the force applied and the distance a load is moved.

work done (J) = force (N) × distance (m)

FIGURE 1.6.4b: Which forces are doing work in these situations?

3. Calculate the work done in the following situations.
   a) A man uses a force of 50 N to push a box 1 m along a smooth floor.
   b) A striker at a fairground uses a force of 100 N to raise the puck a height of 6 m.

4. How much work is done by the engine of a car that applies a force of 20 000 N to move the vehicle 1 kilometre?

6 .4

Force multipliers and energy　》》

In Chapter 5, you learned that a lever is a machine that can multiply force. This means that a large load can be moved by a smaller effort. How is this possible? The answer is that the effort force on the lever acts over a larger distance.

Did you know...?

Just as work is done in making a car move, it is also done in making it stop. In some racing cars so much energy is transferred by the brakes that the brake disks glow.

FIGURE 1.6.4c: Which job is easier?

Look at the girl lifting the stone with a lever in Figure 1.6.4c. She pushes down on the lever with a force F_1.

Work done by girl on lever = $F_1 \times d_1$

Work done on stone by lever = $F_2 \times d_2$

Energy is conserved, so:

work done by girl = work done on stone

$$F_1 \times d_1 = F_2 \times d_2$$

So the stone can be lifted using a force F_1. Without a lever, the girl would have to use a larger force F_2.

FIGURE 1.6.4d

5. a) In Figure 1.6.4d, what is the work done by the man using the lever?
   b) What must the work done be when the box is lifted for energy to be conserved?
   c) If the weight of the box means it needs a force of 200 N to lift it, how high will the box rise up?

Key vocabulary

work done

SEARCH: work done　209

Looking at dynamos

We are learning how to:

- Describe the energy changes in a dynamo.
- Explain how a dynamo works.

A dynamo is a device that can transfer energy of movement into electrical energy. The idea behind it was Michael Faraday's. He built a machine in 1831 that consisted of a copper disc rotating between two magnets. This generated a very small amount of electricity. It paved the way to the generation of electricity on a global scale.

FIGURE 1.6.5a: How does this bike lamp work without a battery?

Uses of dynamos

Have you ever wondered how the lights on some bicycles work without using batteries, or how some torches are able to operate simply by turning a handle? When an electricity supply or batteries are not available, **dynamos** can be very useful to produce small electric currents.

During World War 2, portable radios were operated using dynamos. A handle had to be turned, and the dynamo converted **kinetic energy** into energy of an electrical current, which enabled the radio to work.

Dynamos are not used much in our country today because we have mains electricity from the National Grid, and batteries are widely available. However, dynamos are still used extensively in some countries. Wind-up radios have been designed with built-in torches and mobile phone chargers.

FIGURE 1.6.5b: Dynamos transfer energy of rotary motion to electrical energy.

1. The photos show some uses of dynamos. Can you think of one other device that could be run using a dynamo?

2. Suggest some advantages of using dynamos.

How do dynamos work?

Dynamos rely on **magnetism** and movement to work. If a piece of wire is moved between two magnet poles, the kinetic energy of the wire's movement is transferred as **electrical energy**. If the wire is in a circuit, an electric current is produced.

A dynamo has a coil of wire that moves (or the magnets may move). This increases the amount of electrical energy transferred.

3. What useless energy transfers would there be in a dynamo?

4. Draw an energy transfer diagram for a dynamo.

Increasing the energy

To increase the amount of electrical energy converted by a dynamo, many turns of wire are used on the coil and the coil is turned as fast as possible in a strong magnetic field.

Dynamos used to be operated using water power or steam to spin the coil. These were able to supply plenty of energy by movement. Victorian lighting systems in large houses were run by this system, before the National Grid was established.

Did you know...?

The invention of the dynamo stimulated much interest and research into generating electrical energy. The power that dynamos could produce was limited. Nikola Tesla, one of the greatest inventers, changed the technology to produce a more powerful form of electricity, used today worldwide.

FIGURE 1.6.5c: How could energy from a water wheel supply electrical energy for lights?

5. How can you increase the electrical energy transferred in a dynamo to produce a bigger current?

6. What are the advantages of using water power to provide the kinetic energy for a dynamo?

7. What are the advantages of using steam to power the dynamo?

Key vocabulary

dynamo

kinetic energy

magnetism

electrical energy

Understanding elastic potential energy

We are learning how to:

- Describe different situations that use the energy stored in stretching and compressing elastic materials.
- Describe how elastic potential energy in different materials can be compared.
- Explain how elastic potential energy is transferred.

Elastic materials have the ability to store energy ready for use. The muscle tissue in animals consists of fibres of protein that can expand and contract, providing a potential store of elastic energy. This ability allows us to jump and move – and allows fleas to jump more than a hundred times their own height!

FIGURE 1.6.6a: A flea's jump is an example of elastic potential energy being transferred.

What is elastic potential energy?

Energy is stored when an elastic material is stretched or compressed (squashed) by a force. You do work when you pull an elastic band or squash a spring. This transfers energy, which is stored as **elastic potential energy**.

The stored energy is transferred when the elastic material returns to its original shape.

The further a material is stretched or compressed, and still be able to return to its original position, the more elastic potential energy can be transferred.

1. In which of the situations in Figure 1.6.6b is more elastic potential energy transferred?

2. What causes the jack-in-the-box to bounce up when the lid is opened?

Applications of elastic potential energy

Catapults and archery bows use elastic materials. Elastic potential energy is stored when the elastic is stretched or the bow is bent. More elastic potential energy is stored if the elastic is harder to stretch because more work is done in pulling it back.

FIGURE 1.6.6b: What do these have in common?

Some shock absorbers in cars have strong springs. When driving over a bump, energy is transferred by movement into elastic potential energy in the springs. This energy is released slowly when the car gets beyond the bump.

3. Some students are testing two different elastic materials for use in a catapult. They want to find out which would transfer more energy.

    a) How should they make the investigation a fair test?

    b) What should they measure to collect evidence?

4. Describe the energy transfers in a wind-up clock.

Explaining elastic potential energy

Elastic materials, such as rubber, are made up of molecules that are bound together. When the material is stretched, the bonds between the molecules store potential energy.

In its relaxed state, rubber consists of long strands of molecules which are all coiled up. When the rubber is stretched, the coils become elongated and straightened, enabling the rubber to extend in length. When the stretching force is removed, the molecules return to their coiled-up state and the material returns to its original length.

> ### Did you know…?
>
> Most elastic materials can stretch up to five times their original length. The first type of elastic material was natural rubber, made from the sap of rubber trees. Scientists have recently invented a gel material that can stretch up to 20 times its original length and still recover. It has a possible application as artificial cartilage, because it is also extremely strong.

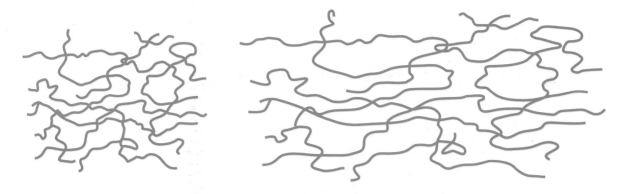

FIGURE 1.6.6c: Molecules of rubber in relaxed and stretched states

The elastic potential energy stored in a rubber band or a spring is equal to the **work done** in stretching it. This energy can be transferred as kinetic energy when the stretching force is removed.

5. Can all materials store elastic potential energy? Explain your answer.

6. How would you test which had more elastic potential energy – a coiled metal spring or an elastic band?

Key vocabulary

elastic potential energy

work done

Knowing the difference between heat and temperature

We are learning how to:

- Recognise what we mean by temperature.
- Describe how temperature differences lead to energy transfer.
- Explain the difference between heat and temperature.

The hottest temperature ever recorded on Earth was 56.8 °C in Death Valley, USA in 1913. The coldest was –89 °C in Vostock, Antartica in 1983. What do we actually mean by how hot or how cold something is?

Temperature

We use a scale of **temperature** to measure how hot or how cold something is. The common unit is **degrees Celsius** (°C). The instrument we use to measure temperature is called a thermometer. Standard thermometers measure temperatures from 0 °C (the temperature at which water freezes) to 100 °C (the temperature at which water boils).

1. Put the following objects in order of temperature, with the hottest first.

A the Sun's surface, B boiling water, C volcano lava, D glacier of ice

2. Using Figure 1.6.7a to help you, match up the objects and their corresponding temperatures in Table 1.6.7.

TABLE 1.6.7

Object	Temperature (°C)
1 body temperature	a) 20
2 bath water	b) 5
3 temperature of a hot sunny day	c) 57
4 highest air temperature recorded	d) 30
5 room temperature	e) 37
6 cold drink from the fridge	f) 50

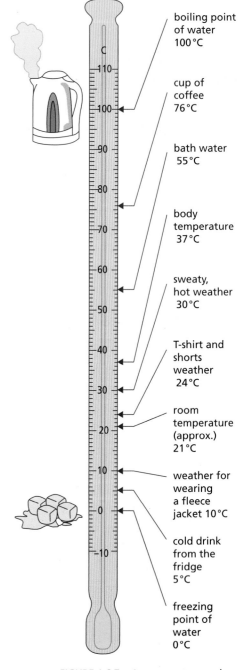

boiling point of water 100 °C

cup of coffee 76 °C

bath water 55 °C

body temperature 37 °C

sweaty, hot weather 30 °C

T-shirt and shorts weather 24 °C

room temperature (approx.) 21 °C

weather for wearing a fleece jacket 10 °C

cold drink from the fridge 5 °C

freezing point of water 0 °C

FIGURE 1.6.7a: A temperature scale

Heat flow

If there is a difference in temperature between two objects in contact, or between an object and its surroundings, there is a transfer of energy as a flow of **heat**. Heat always flows from the hotter object to the colder object. Heat will continue to flow in this direction until the two objects reach the same temperature. The greater the difference in temperature, the faster the flow of heat.

We use this principle of heat transfer to heat and cool objects or our surroundings. If you put some food at room temperature (20 °C) in a fridge (at 5 °C), energy from the warmer food will transfer to the colder surroundings of the fridge. This will reduce the temperature of the food and cool it down.

> **Did you know...?**
>
> The coldest temperature possible is called 'absolute zero'. It is −273.15 °C. Some scientists have won the Nobel prize for finding ways to cool matter to within billionths of a degree of absolute zero.

3. Look at the diagrams in Figure 1.6.7b. State the direction of heat transfer in each situation.

4. Put the diagrams in order of the quickest to transfer heat to the slowest. Explain your answer.

5. If you wanted to cool a container of water as quickly as possible, would you put it in a fridge or in a freezer? Explain your answer.

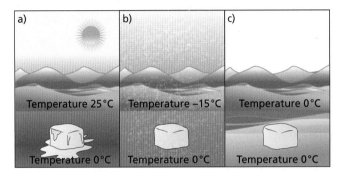

FIGURE 1.6.7b: Will the heat flow from the ice cube or to the ice cube?

Heat and temperature

Think about a cup of hot water at 80 °C and a big bucket of warm water at 30 °C. You know that the water in the cup has a higher temperature. But which holds more heat energy?

Say that a cubic centimetre of water particles in the bucket has 40 J of energy, and a cubic centimetre of water particles in the cup has 250 J. Which would hold the most energy overall – the cup or the bucket?

- Temperature is a measure of the energy of the individual particles – how fast they are moving about.

- Heat is the total energy of all the particles – a measure of not only how fast particles are moving but also the total number and type of particles.

6. Which would have most energy – a cup of boiling water or a gigantic iceberg? Explain your answer.

7. Explain in terms of the particle model why the direction of energy flow is always from a hotter to a cooler object.

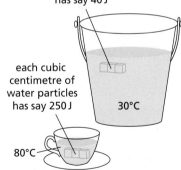

FIGURE 1.6.7c: The water in the cup has a higher temperature. But which has more energy?

Key vocabulary

temperature

degrees Celsius

heat

Thinking about fuels

We are learning how to:

- Identify examples of fuels and their uses.
- Describe the combustion of fuels and recognise that different fuels transfer different amounts of energy.
- Describe the advantages and disadvantages of using different fuels.

Dried animal dung has served as a fuel for humans, all over the world, since prehistoric times. Today, over two billion people still use this as their main source of fuel. Dung from cows, buffalo and camels is commonly heated and burned. The energy is used for cooking food, heating and drying.

FIGURE 1.6.8a: What are the advantages of using animal dung as a fuel?

Common fuels

Fuels are chemicals that transfer energy to the surroundings by heat when they burn. They have **chemical potential energy**. Many different types of fuels exist, and they may be solids, liquids or gases.

The most commonly used fuels are coal, crude oil and methane (natural gas). Car engines use liquid fuels such as petrol and diesel, or a gas called propane. Aeroplanes use a heavier liquid fuel called kerosene, and racing cars use a lighter liquid fuel such as nitromethane. All of these are processed from crude oil. They have the advantage of burning quickly and easily, transferring energy rapidly.

Wood and wax are solid fuels that provide heat and light. These burn much more slowly than liquid fuels.

TABLE 1.6.8: Some common fuels

Fuel	Solid, liquid or gas	Uses
coal	solid	making electricity; coal fires
natural gas	gas	making electricity; cookers
petrol	liquid	fuel for vehicles
diesel	liquid	in engines of cars, boats and trains
kerosene	liquid	fuels for planes
methanol and nitromethane	liquid	fuels for racing cars
wood	solid	fires and boilers

1. State one advantage of using a liquid fuel, rather than a solid fuel.

2. Suggest how life would change if we ran out of crude oil.

Did you know...?

Hydrogen is being developed as an alternative fuel to hydrocarbons. Water vapour is given off when hydrogen burns. This returns back to the Earth as rain.

Combustion >>>

When fuels are burned, they combine chemically with **oxygen** from the air. This process releases heat energy – it is called **combustion**. A heat source, for example a match or a spark in an engine, starts the process off. Chemical energy is transferred to the surroundings by heat, and also by light and a small amount by sound.

3. Describe the energy transfer when a fuel is burned. State the useful and useless energy transfers.

4. How might fuels vary from each other?

FIGURE 1.6.8b: Burning hydrocarbons causes global problems. How do these occur?

Problems with fuels >>>

Fuels derived from crude oil belong to a family of chemicals known as **hydrocarbons**. When they burn, the following chemical reaction takes place, transferring energy by heat and light to the surroundings:

hydrocarbon + oxygen → carbon dioxide + water

Carbon dioxide is a greenhouse gas and so contributes to global warming. Burning hydrocarbons also produces acid rain. This damages forests, plants and living things in lakes.

We have used so much crude oil that the Earth's supply of oil is running low. Crude oil is a **fossil fuel**, meaning it takes millions of years to form.

5. When hydrogen is used as a fuel, the only product is water. What advantage does this fuel have compared with hydrocarbon fuels?

6. Because hydrocarbons from fossil fuels cause so many problems, why are we still using them?

Key vocabulary

chemical potential energy

oxygen

combustion

hydrocarbon

fossil fuel

Investigating fuels

We are learning how to:

- Describe how to measure the energy of fuels.
- Collect evidence to investigate the energy of different fuels.
- Present data using appropriate graphs and evaluate the quality of evidence collected.

Fuels vary widely in their properties. Hydrogen and natural gas are extremely explosive; petrol is highly flammable; lumps of coal and logs of wood are quite hard to set alight. Fuels also vary in how much energy they can transfer per gram. Hydrogen transfers the highest amount of energy, whereas wood transfers the least.

Energy in fuels 》

The amount of energy transferred by the **combustion** (burning) of a fuel depends on:

- the amount of fuel burnt
- how much energy is stored in the fuel.

The more fuel burned, the greater the amount of chemical energy transferred by heat. The type of fuel will also determine how much energy is transferred. With some fuels, such as petrol, only a small amount has to be burned to obtain a large amount of heat. In contrast, a lot of wood has to burn to obtain the same amount of heat.

1. Name two fuels that you use regularly.

2. If you wanted to transfer more heat from a fuel, what two things could you do?

Differences between fuels 》》

Fuels with the most chemical energy to transfer will produce the highest amount of heat, using the least amount of fuel.

One way to compare the energy transferred from different fuels is to burn the same mass of each fuel, use the heat energy to warm the same amount of water and record the temperature rise of the water. Fuels that transfer the most energy will produce the highest rise in temperature.

FIGURE 1.6.9a: How are fuels different?

3. Which fuel in Table 1.6.9 has the lowest energy per gram?

4. Approximately how much of the fuel you have named in question 1 would be needed, to give the same energy as 1 g of coal?

5. Use the information in the table to suggest which fuel might be suitable for use in a rocket. Give a reason for your answer.

TABLE 1.6.9: Energy of different fuels

Fuel	Energy (kJ/g)
hydrogen	143
petrol	46
diesel	45
aviation fuel	43
paraffin	42
natural gas	37
coal	33
wood	12

Interpreting graphical data ⟩⟩⟩

Graphs are a useful way to compare energy data for different fuels. Different types of graph show data in different ways. Look at those in Figure 1.6.9b, which present the data from Table 1.6.9.

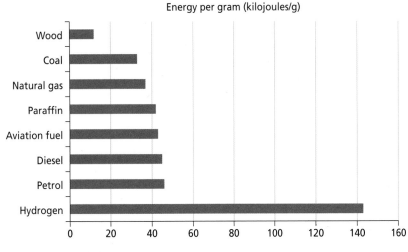

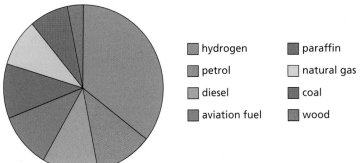

FIGURE 1.6.9b: Two different types of graph showing the energy of different fuels

6. Compare and contrast the two graphs in Figure 1.6.9b. What are their strengths and their weaknesses?

7. Why is it not possible to produce a line graph from this data?

8. Can you deduce any patterns from these graphs?

Did you know...?

Uranium is the fuel with the highest amount of energy per gram on Earth. It does not burn but undergoes nuclear reactions to transfer energy by heat. One kilogram of uranium will release the same amount of energy as 3000 tonnes of coal!

Key vocabulary

combustion

Applying key ideas

You have now met a number of important ideas in this chapter. This activity is an opportunity for you to apply them, just as scientists do. Read the text first, then have a go at the tasks. The first few are fairly easy, then they get a bit more challenging.

Energy changes – making electricity

The dynamo was the first device to transfer energy of movement to the energy of an electric current. In a dynamo there is a magnet and a coil of copper wire. When the coil spins in the magnetic field, energy is transferred to the wires as electricity. It also works if the magnet spins, with the coil stationary.

The kinetic energy needed to operate a dynamo can come from different sources. Originally a hand-turned wheel was used. Later designs incorporated gears. Eventually bicycles were used. The dynamo head was placed against the wheel of the bicycle.

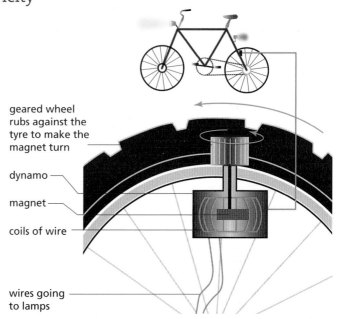

geared wheel rubs against the tyre to make the magnet turn

dynamo

magnet

coils of wire

wires going to lamps

FIGURE 1.6.10a: Bicycle dynamo

The spinning wheel caused the head to turn, transferring energy from the movement of the bike into electrical energy. This method is still used to operate the lights on a bike at night, without the need for batteries. In larger applications, dynamos were driven by water wheels, using moving water from a river or waterfall. The energy of the moving water provided the energy needed to spin the dynamo and produce electricity.

The design of dynamos has changed over time. Figure 1.6.10a shows a modern design, fitted with gears. Gearing in a dynamo means that fewer turns are needed to produce the same amount of electricity.

Dynamos are only used nowadays for applications where mains electricity is unavailable – for example camping torches, radios, mobile phone chargers and emergency lights.

Electrical energy from chemical energy

Most of our electricity comes from burning a fuel such as gas, coal or oil. These are all fossil fuels. The stored chemical energy is transferred as heat to water, turning it into steam. The steam has kinetic energy and drives a turbine, which is rather like a water wheel. This is used to turn a generator, a similar device to a dynamo, and so produce electricity. Most of the electricity in the world comes from this process. The generator is more efficient than a dynamo and more electricity can be made from it.

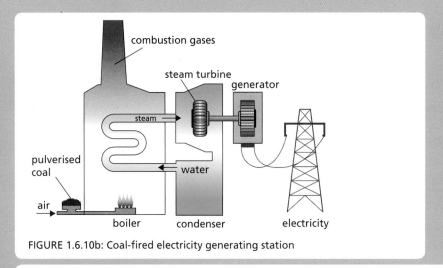

FIGURE 1.6.10b: Coal-fired electricity generating station

Task 1: Identifying energy changes

Explain how energy is transferred in a water-fed dynamo, like that in Figure 1.6.5c.

Task 2: Useful energy changes

Read the text and use the information to draw energy transfer diagrams to represent the following:

a) a bicycle dynamo making a light work

b) a hand-held dynamo operating a radio, like that in Figure 1.6.5b.

Task 3: Useless energy

Think about a bicycle dynamo. What are the ways in which energy is transferred that are not useful?

Task 4: Heat and temperature

Consider these two different systems that make electricity.

The first is a coal-fired power station. The burning coal turns water into steam to drive a turbine generator. The second is a hydroelectric power station, in which a waterfall drives a water wheel powering a dynamo. Compare and contrast the energy transfers of the two systems.

Task 5: Different processes to transfer energy

Using a hand-crank dynamo to light a bulb uses muscles, which contain chemical energy and elastic potential energy. Explain each process of the energy transfer, including useful and useless transfers. Make justified predictions about the efficiency of each process and draw a Sankey diagram for the overall energy transfer.

Exploring sound

We are learning how to:
- Identify how sounds are made.
- Describe how sound waves transfer energy.
- Explain how loud and quiet sounds are made.

Sound is hugely important throughout the animal kingdom as a means of communication, location and defence.

Making sounds

If you place a finger over your voice box when speaking or singing, you will feel your voice box **vibrate**. This is where the sound comes from.

When an instrument is plucked or blown through, the string or the air vibrates. Often the vibrations are too small to see.

All vibrations result in a sound. The vibrations from the object are passed on to air particles. These air particles bump into others and the wave progresses. Eventually the energy of the vibrations is transferred to your ears.

FIGURE 1.6.11a: How does a guitar make a sound?

1. What causes the sound when a bell is rung?

2. How does the sound from a concert reach the audience?

Making waves

Energy is transferred by sound in the form of waves. In Figure 1.6.11b a slinky spring provides a model that shows how these waves work. When you push the end of a slinky back and forth, some of the coils squash together and others pull apart. A wave of energy passes along the length of the spring. A wave like this which travels in the same line as the vibrations of the source is called a **longitudinal wave**.

source moves left and right

coils move left and right

energy transfer

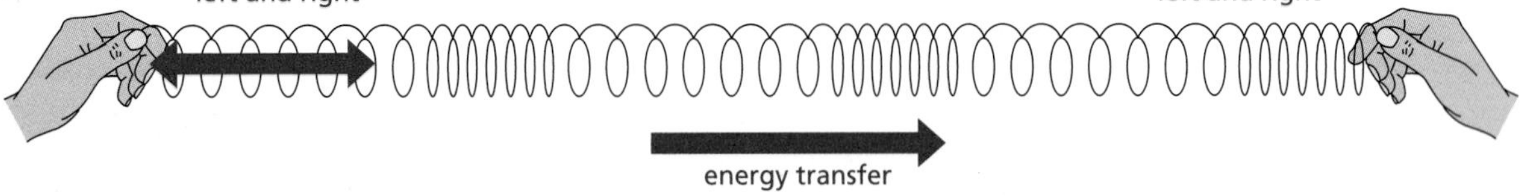

FIGURE 1.6.11b: Why is this called a longitudinal wave?

A sound wave works in the same way. Vibrations push air particles together and also pull them apart, creating a longitudinal wave of energy. The energy passes from the source of the vibration to our ears.

3. Describe the movement of air particles in a longitudinal sound wave.

4. What happens to the energy in a longitudinal wave?

Louder and quieter sounds

Sounds can be made louder by increasing the energy in the vibration. Plucking a string harder, blowing harder through a wind instrument or beating a drum harder will all transfer more energy. The loudness of sound is measured in a unit called a **decibel (dB)**. The loudest sound that humans can listen to without damage to their hearing is about 120 dB.

The size of a vibration is represented by its **amplitude**. Figure 1.6.11d shows that the amplitude is the maximum distance that a particle travels in the to-and-fro vibration. The greater the amplitude, the greater the energy of the vibration and the louder the sound. In other words, a bigger wave will transfer more energy and be heard as a louder sound.

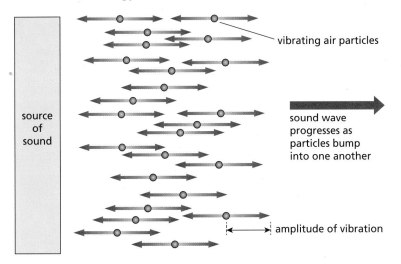

vibrating air particles

source of sound

sound wave progresses as particles bump into one another

amplitude of vibration

FIGURE 1.6.11d: What effect will a smaller amplitude have?

5. Look at Table 1.6.11. Match the sounds to the correct loudness.

6. The loudness of a sound also depends on the distance from the source. Explain what happens to the energy as you get further away.

7. Is there a limit to how loud a sound can be made? Explain your answer.

Did you know...?

The ocean-dwelling tiger pistol shrimp is known to produce the loudest sound on Earth, reaching over 200 dB. It uses the sound as a defence mechanism. The vibrations can kill prey and fish up to 2 metres away!

FIGURE 1.6.11c: A pistol shrimp

TABLE 1.6.11

Sound	Loudness (dB)
1 whisper	a) 80
2 phone dial tone	b) 140
3 jet engine	c) 100
4 motorbike	d) 30

Key vocabulary

vibrate

longitudinal wave

decibel (dB)

amplitude

Describing sound

We are learning how to:

- Describe how the pitch of a sound wave can be changed.
- Apply the terms frequency, wavelength and amplitude to different waveforms.

There are many different types of sound. Think of the sounds made by a whale compared with the high-pitched screeching of a monkey, or the sound of a bass guitar compared with a violin. Differences in sound waves arise from different characteristics of the sound waves.

FIGURE 1.6.12a: How would you describe the scream of a monkey?

What is pitch? »

A ship's horn produces a sound that is very deep and low – this is known as a low pitch. Whistles, alarms and sirens produce high-pitched sound.

The pitch of a note is also called its **frequency**. A high frequency means a high pitch and a low frequency means a low pitch. Muscial notes change in pitch by changing the frequency of the vibration.

Feel your voice box as you make sounds of different pitches. What do you notice?

1. Describe one other sound with a low pitch and one other sound with a high pitch.

2. What is meant by the 'frequency' of a note?

Frequency, wavelength and amplitude »

Sound waves can be represented in a diagram like that shown in Figure 1.6.12b. The curve, or **waveform**, is a graph of the displacement of the air particles at different distances along the wave. The **wavelength** is the distance along a wave from one point to the next point where the wave motion begins to repeat itself.

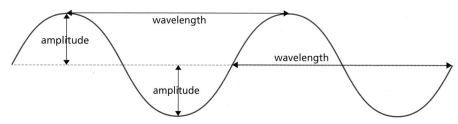

FIGURE 1.6.12b: Parts of a wave

The higher the frequency of a wave, the shorter the wavelength. A high frequency means more vibrations are produced per second.

The maximum dispacement is called the **amplitude**. The energy transferred by the wave depends on this. The larger the amplitude of a sound wave, the louder the sound.

3. Why is it more useful to use the wave representation in Figure 1.6.12b, compared with a drawing of a longitudinal vibration, as in Figure 1.6.11d of Topic 6.11?

4. How could you tell from a waveform whether a sound is getting:

   a) louder?   **b)** higher pitched?

Interpreting sound waves

All sound waves can be detected using a **microphone** and shown as a waveform on a screen. The microphone receives the sound waves and converts them into electrical signals. Some typical examples are shown in Figure 1.6.12c.

Detectives use traces like these to match voices patterns from recordings to known criminals or to identify patterns from the shots of particular guns.

5. **a)** Which wave in Figure 1.6.12c results from the loudest sound?

   b) Which wave results from highest-pitch sound?

   c) Which wave is transferring the most energy? Explain your answer.

6. Draw waves to represent a loud high-pitched sound and a quiet low-pitched sound.

7. Look at the graph in Figure 1.6.12d, which shows the sound wave detected from a gun as time progresses. Describe what is happening to the frequency, wavelength and amplitude of the wave.

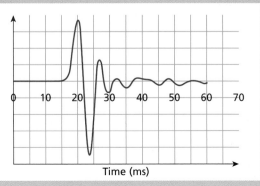

FIGURE 1.6.12d: Sound wave from a gun

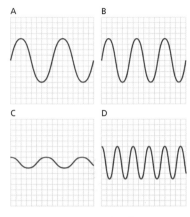

FIGURE 1.6.12c: How are these waves different?

Did you know...?

Microphones have a thin diaphragm made of plastic or metal. This vibrates when even small sound vibrations hit it. The energy from the vibrations is transferred by electric current and can be fed to a loudspeaker. This transfers the electrical energy back into sound energy.

Key vocabulary

frequency

waveform

wavelength

amplitude

microphone

Measuring the speed of sound

We are learning how to:

- Describe what an echo is.
- Describe how the speed of sound can be measured using echoes.
- Calculate distances using ideas about echoes.

Echoes are used by bats and dolphins to navigate, find prey and find mates. Dolphins can tell the difference between a golf ball and a ping-pong ball using echoes, and bats can detect tiny insects in pitch-black darkness. Humans have learned from these creatures to use echoes in similar ways.

FIGURE 1.6.13a: Some bats are completely blind. How do they find food?

What is an echo?

Sound waves can bounce back from a surface. We call this **reflection** of sound an **echo**. A hard surface will reflect more sound than a soft surface, making a stronger echo. Shouting across a rocky outcrop or in a long tunnel can produce good echoes.

An echo transfers less energy back to the listener, and so is quieter than the original sound. This is because some energy has been transferred as heat in the material it is reflected from.

1. What do we mean by a 'reflection' of sound?

2. Which types of place would not allow echoes to be created?

FIGURE 1.6.13b: Why is this a particularly good place for echoes?

Measuring the speed of sound

An echo is the reflection of a sound wave from a surface, including back towards the sender. The echo travels twice the distance from the sender to the surface. When the echo from a sound wave comes back, it can be used to calculate the **speed of sound**.

$$\text{speed of sound} = \frac{\text{total distance travelled by the echo (m)}}{\text{time taken (s)}}$$

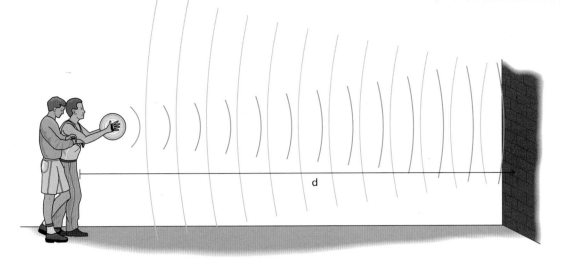

FIGURE 1.6.13c: How to measure the speed of sound

3. Describe how echoes can be used in this way to measure the speed of sound.

4. A man fires an air pistol towards a wall, a distance of 1500 m from him. It takes 10 seconds before he hears the echo. Calculate the speed of sound from this information.

5. How far away is a quarry if the sound of blasting is heard 3.5 seconds after the explosion?

   (distance = speed × time; the speed of sound in air is 330 m/s)

Sonar

Naval ships use **sonar** to locate hidden submarines. They send sound waves through the water from the ship. The sound waves reflect off the sea floor and return to the ship as an echo. If a submarine is located under the passing ship, the echo returns more quickly, because the sound has a shorter distance to travel.

Fishermen also use this method to locate shoals of fish. The speed of sound through water is 1500 m/s. When calculating the distance of an object from the boat, remember that the sound waves travel to the sea bed and back again:

distance of object to boat = ½ × speed of sound × time taken

6. A boat uses its sonar system to send a sound signal directly downwards to the sea bed. An echo is recorded on the boat 2 seconds later. How deep is the water?

7. A battleship records a sonar echo from a submarine below from the ship. There is 5 s between sending the sonar signal and receiving the echo. How far away is the submarine?

> **Did you know...?**
>
> Some blind people learn to use echoes as a way of locating objects around them. By clicking their tongue and listening to the echoes, they can avoid walking into obstacles.

Key vocabulary

reflection

echo

speed of sound

sonar

Understanding how sound travels through materials

We are learning how to:

- Recognise how the speed of sound changes in different substances.
- Use the particle model to explain why there are differences when sound travels through solids, liquids and gases.

Whales are known to transmit sounds in the ocean over distances of 700 km. If whales were to transmit these same sounds in the air, would they travel faster or slower?

FIGURE 1.6.14a: How do sounds from whales travel under water?

Sound in a vacuum

Most of the sounds that you hear are transmitted by vibrating air **particles** (particles of gas). Sounds can also travel through solids and liquids. Sound waves need particles of matter to transmit energy. As the particles vibrate, the energy is passed on to adjacent particles and carried in the form of a wave.

Sounds cannot travel through a **vacuum**, nor through space, which has hardly any particles in it.

1. Why can sound travel not through a vacuum?

2. How is it possible for sounds to travel through solids?

Speed of sound through different materials

Table 1.6.14 shows the speed that sound travels through different materials.

3. a) In which material does sound travel the fastest?

   b) In which material does sound travel the slowest?

   c) Does sound travel fastest in solids, liquids or gases?

TABLE 1.6.14: Speed of sound in different materials

Material	Speed of sound (m/s)
air	343
oxygen	316
carbon dioxide	259
water	1482
lead	1960
copper	5010
steel	5960
diamond	12 000

Particles of matter in solids, liquids and gases differ in their arrangement and behaviour. This affects how well sound waves can travel through them. The speed at which the wave moves depends on the arrangement of the particles, the elastic nature of the forces between them, and how fast the particles are moving.

- In a gas the particles are very far apart. Sound travels slowly because the particles do not collide very often.

- In a liquid the particles are much closer to one another. Sound travels more quickly because the particles are able to collide with each other much more frequently. Sound travels about five times faster through liquids than it does through gases.

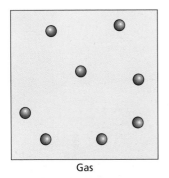
Gas

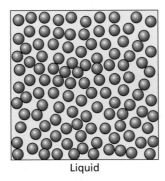
Liquid

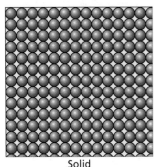
Solid

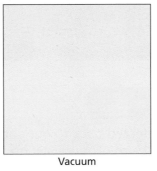
Vacuum

> **Did you know...?**
>
> Native Americans used to put an ear to railway tracks to know when trains were coming. This is a dangerous thing to do because you can never tell how soon the train will arrive.

FIGURE 1.6.14b: The particle theory of matter explains how sound travels in a solid, liquid and gas. Why does sound not pass through a vacuum?

- In a solid the particles are packed very closely together. Also, the forces between the particles are more elastic. The vibrating particles collide with neighbouring particles and bounce back very quickly, so the sound wave progresses very quickly.

4. Do you think sound will travel faster through water or ice? Explain your answer.

5. Why do you think sound travels much faster through some solids compared to others?

6. Temperature can also affect the speed of sound. Develop a **hypothesis** to explain why.

Key vocabulary

particle

vacuum

hypothesis

Learning about the reflection and absorption of sound

We are learning how to:

- Recognise which materials affect the quality of sound.
- Analyse the effect of different materials on sound waves.
- Use ideas about energy transfer to explain how soundproofing works.

Concert halls are designed for good acoustics – so that the music sounds good to the whole audience. This means controlling the amount of echo and making sure sound reaches all corners. Different materials and shapes are used to achieve this.

FIGURE 1.6.15a: Can you identify materials that help to reflect sound and those that help to absorb it?

Effect of materials on sound waves

Echoes are sound waves that are reflected back to our ears. Hard, flat surfaces **reflect** sound well and produce strong echoes.

Soft surface materials that contain lots of air pockets, like fabric, foam and sponge, are not good at reflecting sound. They **absorb** it. The sound waves transfer energy to the air in the pockets so less is reflected.

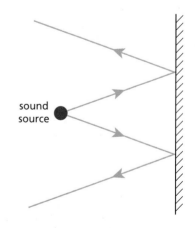

sound source

FIGURE 1.6.15b: What kind of echoes will this produce? How can the echoes be reduced?

1. What do we mean by 'absorbing' sound?
2. a) What would you hear if the sound waves from a bell were directed at a metal panel?
   b) What would you hear if the sound waves from a bell were directed at a panel made of sponge?
3. When might it be useful to absorb sound waves?

Some materials can be shaped to reflect sounds in different ways. Look at the jagged surface in Figure 1.6.15c. When sound waves hit this surface, the reflected waves do not bounce back to the source. They are, instead reflected randomly, mostly away from the source.

The curved surface, on the other hand, reflects the sound until all the energy focuses towards a particular point. The sound at this point will be the loudest, whereas in places away from it, hardly any sound will be heard at all.

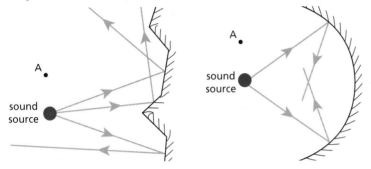

FIGURE 1.6.15c: How sound is reflected off a jagged and a curved surface

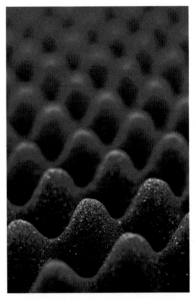

FIGURE 1.6.15d: This material is used in soundproofing. What makes it a good choice?

> **4.** Imagine you are standing in position A in each of the diagrams in Figure 1.6.15c. Describe what you will hear if the surface is:
>
> **a)** jagged   **b)** curved
>
> compared with the flat surface in Figure 1.6.15b.

Soundproofing ▶▶▶

When sound waves hit soft surfaces, they are absorbed by the air pockets. The sound waves become trapped, bouncing around in the air pockets, until all the energy is transferred as heat. Any sound reflected from the surface is therefore much quieter, as the sound waves have much less energy.

These soft materials are useful as **soundproofing**. A vacuum is also useful in soundproofing. Sheets of glass with a near-vacuum between them (very few gas particles) are very effective in stopping sound.

In the outdoor environment, trees, embankments and dense bushes are often for soundproofing around mining areas.

> **5.** What happens to the energy of the sound wave during absorption?
>
> **6.** Design a soundproofing plan for a hospital in a busy town centre.

Did you know...?

A 'whispering gallery' is the name given to a large circular room, where a whisper made in one place is reflected to the opposite side of the room and heard there but nowhere else. St Paul's Cathedral contains one.

Key vocabulary

reflect

absorb

soundproofing

Hearing sounds

We are learning how to:

- Describe the structure and function of different parts of the ear.
- Explain how the ear is able to hear and detect sounds.

The ability to hear is important in all animals for communication, hearing predators, knowing when there is danger and seeking prey. The human ear relies on a combination of processes and ingenious engineering to help us to identify the wide range of sound waves we receive.

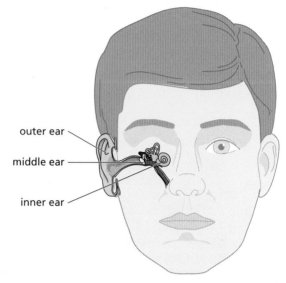

FIGURE 1.6.16a: The human ear is not just the part you can see.

Ears for hearing

A human ear is divided into three parts – the outer ear, the middle ear and the inner ear.

The outer ear is the part that can be seen, on the outside of your head. Its job is to capture sound waves. The waves pass along the **ear canal** to the **ear drum**. This separates the outer ear and middle ear. The ear drum transfers the energy from the vibrations to bones called **ossicles** in the middle ear, which make the tiny vibrations much bigger. This energy is passed on to the inner ear, which contains specialised cells that detect the vibrations and convert them into electrical signals. These are sent to the brain, which interprets them.

1. What are the jobs of the outer, middle and inner ear?

2. Suggest why are our ears are located in our head.

Structure of the human ear

The function (job) of the ear is to transfer energy by sound into electrical impulses that are interpreted by the brain.

Figure 1.6.16b shows the detailed structure of the ear, and Figure 1.6.16c describes what happens to the sound waves as they enter the ear.

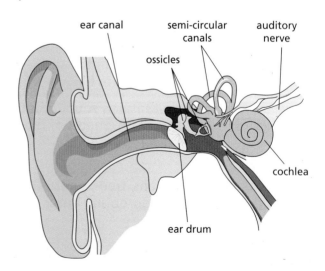

FIGURE 1.6.16b: Parts of the ear

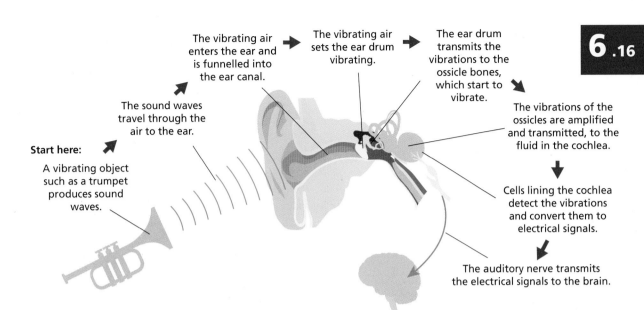

The vibrating air enters the ear and is funnelled into the ear canal.

The vibrating air sets the ear drum vibrating.

The ear drum transmits the vibrations to the ossicle bones, which start to vibrate.

The sound waves travel through the air to the ear.

The vibrations of the ossicles are amplified and transmitted, to the fluid in the cochlea.

Start here:
A vibrating object such as a trumpet produces sound waves.

Cells lining the cochlea detect the vibrations and convert them to electrical signals.

The auditory nerve transmits the electrical signals to the brain.

FIGURE 1.6.16c: How we hear

3. Suggest why incoming sound vibrations need to be amplified (amplitude made bigger) in the ear.

4. Where in the ear are:

  a) electrical signals transmitted to the brain?

  b) sound vibrations amplified?

  c) vibrations first detected?

Adaptations of the ear >>>>

The ear drum is like a tiny drum skin. Muscles keep it very rigid, so even the slightest vibration causes it to move back and forth.

The three small, connected ossicles are called the malleus, incus and stapes. The malleus is connected to the ear drum, and the stapes is connected to the **cochlea**.

The ear drum has a surface area of about 55 mm², but that of the stapes is only about 3 mm². The energy of the vibrations is transmitted through a much reduced area. This multiplies the pressure by about 20 times, so amplifying the vibrations as they are passed on to the cochlea.

The cochlea is filled with fluid, enabling the the sound vibrations to travel much faster. It has thousands of tiny hair cells that convert the sound wave to electrical signals, which are passed on to the **auditory nerve** and to the brain.

5. What is the result of the difference in surface area in the middle ear?

> **Did you know...?**
>
> The cochlea is also responsible for detecting and controlling balance.

Key vocabulary

ear canal

ear drum

ossicles

cochlea

auditory nerve

Understanding factors affecting hearing

We are learning how to:

- Describe factors which affect hearing.
- Explain how to prevent damage to ears.
- Understand the term hearing range.

The ability to hear different sounds varies widely across the animal kingdom. Many animals can hear sounds that humans are totally unaware of. Some animals do not have ears, but other organs have adapted to detect sounds. Our own hearing can be affected by factors such as disease and ear damage caused by loud sounds. It is important to learn how to protect our hearing.

FIGURE 1.6.17a: This bushbaby is active only in darkness. It has extremely good hearing.

Protecting our hearing »

The human ear is only able to withstand sounds of a certain **loudness** – too loud and our ears can be permanently damaged. We can do several things to protect ourselves from loud noise:

- turn down the volume of sound-making devices
- increase the distance from the source of the noise
- reduce the time of exposure to loud sounds
- wear **ear defenders**
- obey laws that limit noise in the workplace
- use soundproofing materials.

1. Why is it important to protect your ears?

2. You are the manager of a new quarry that will blast stone from a rock face. Explain what measures you will take to protect the hearing of your workers.

Factors affecting hearing »»

Several factors can affect the health of our ears. Read about these in Table 1.6.17.

3. Why can some ear problems not be cured?

4. Who is most likely to be most at risk of having problems with poor hearing?

TABLE 1.6.17 Causes of ear damage and what can be done

Causes of poor hearing or ear damage	Possible solutions
Ear canal can become blocked with wax.	Have the ear canal cleaned out.
Very loud sounds can rupture the ear drum.	Ear drum may heal itself over a long period of time.
Ear drum can be damaged by infection.	Use antibiotics to get rid of the infection.
Ossicles can become fused together.	An operation is needed.
Infection may occur in the middle ear.	Use antibiotics to get rid of the infection.
Hair cells and nerves in the cochlea may be damaged by loud noises.	There is no cure.
In older people, nerves cells may deteriorate.	There is no cure.

Did you know...?

Elephants can hear frequencies 20 times lower than the lowest frequency we can hear. They use their trunks as well as their ears to detect low frequency vibrations. This enables them to hear other elephants up to 6 km away.

Sound frequencies heard by different animals

The **frequency** of a wave is the number of waves per second. It is measured in hertz (Hz). If a sound wave has a high frequency it means that more vibrations arrive every second – it sounds higher pitched.

We can only hear certain frequencies of sounds. The sounds that we can hear, from the lowest frequency to the highest frequency, is known as our **hearing range**. This range differs widely for different animals – many animals can hear sounds so high pitched or so low pitched that we are unable to hear them. Figure 1.6.17b shows the hearing range for some different animals.

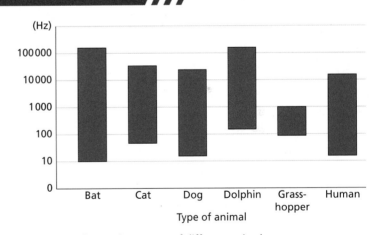

FIGURE 1.6.17b: Hearing ranges of different animals.

5. What is the human hearing range?

6. Draw two diagrams to show what is meant by 'wave frequency'.

7. What range of frequencies can be heard by all the animals in Figure 1.6.17b?

Key vocabulary

loudness

ear defenders

frequency

hearing range

Finding out about sounds we cannot hear

We are learning how to:

- Recognise what is meant by ultrasound and infrasound.
- Describe some applications for ultrasound and infrasound.
- Explain how some applications work.

There are many sounds that humans cannot hear. Infrasounds are sounds at frequencies below our hearing range, and ultrasounds are those above our hearing range. The animal world uses both infrasound and ultrasound for communication and detection. We have also found useful applications for these sound waves that we cannot hear.

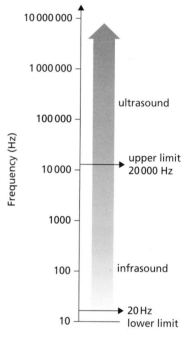

FIGURE 1.6.18a: Range of sound frequencies

What are ultrasound and infrasound?

Humans can hear sounds within the frequency range of 20 Hz to 20 000 Hz.

- Sound below 20 Hz is called **infrasound**.
- Sound above 20 000 Hz is called **ultrasound**.

Humans cannot detect these sounds because our ears are not sensitive enough.

1. Refer back to the 'Did you know…?' box in Topic 6.17.

   a) What is the lowest frequency that elephants can hear?

   b) What do we call that type of sound?

   c) Why can humans not hear such sounds?

Using ultrasound and infrasound

Bats, whales and dolphins emit ultrasound waves in the echo-location of prey, predators and obstacles.

We can use devices that generate ultrasound waves, and also to detect reflected ultrasound. This has led to many useful applications.

- Ultrasound can be used to safely **scan** body organs and unborn babies, allowing checks for anything unusual.
- Metals in aircraft parts and underground pipes can be scanned for cracks using ultrasound.

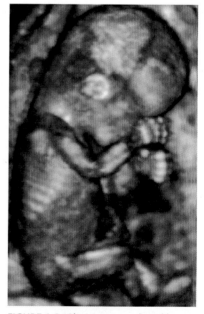

FIGURE 1.6.18b: Image produced by a 3D ultrasound scan of an unborn baby.

- Sonar involves sending ultrasound through water and detecting its reflection from objects.

The high frequency of ultrasound waves means that they transfer energy rapidly.

- Kidney stones can be broken up using ultrasound, without the need for surgery.

- Surgical equipment, electronic components, machinery, jewellery and teeth can all be safely cleaned using ultrasound.

Infrasound waves transfer little energy, but they can be detected by microphones.

- Some large animals use infrasound waves to communicate. Scientists can detect these waves to track herds, in conservation and protection projects.

- Scientists can detect infrasound from volcanoes that are about to erupt, and so warn people of impending risk.

FIGURE 1.6.18c: Ultrasound was used to clean these gear teeth.

- Infrasound can be used to track the passage of meteors in space, so preparing us for any probable collision.

2. Describe two applications of ultrasound or infrasound that can potentially save lives.

3. Why can items such as electronic components not be cleaned using sound waves that we can hear?

How ultrasound scanners work ⟩⟩⟩

An ultrasound scanner contains a special quartz crystal. When a changing electric current is passed through the crystal, it vibrates at very high frequency and emits ultrasound waves.

The ultrasound is directed at the object to be scanned. Different parts of the object reflect the ultrasound by different amounts.

When the reflected ultrasound waves hit the same crystal, it produces a varying electric current that can be detected. This can be built up into a picture using a computer.

4. What are the energy transfers in an ultrasound scanner?

5. You cannot hear ultrasound waves, so how can you be sure they are there?

Did you know...?

Some animals – such as giraffes, whales, hippopotamuses and rhinoceroses – produce infrasound waves to communicate with each other. Elephants produce infrasound waves with their feet. The ground can carry these waves several kilometres to neighbouring herds.

Key vocabulary

infrasound

ultrasound

scan

Checking your progress

To make good progress in understanding science you need to focus on these ideas and skills.

- Recognise that energy is transferred by a range of different processes.

- Interpret and draw energy transfer diagrams for a range of different energy transfers, including gravitational potential energy, elastic potential energy, chemical energy and electrical energy.

- Use Sankey diagrams to explain a range of energy changes and demonstrate that all energy is always accounted for.

- Identify simple energy transfers that involve gravitational potential energy, elastic potential energy and chemical energy.

- Explain how energy is transferred using elastic, gravitational and chemical potential energy.

- Analyse changes in gravitational potential energy in different situations, and compare the energy per gram of different fuels.

- Recognise that work can be done by a force, and that the work done is equal to the energy transferred.

- Calculate the work done in different situations, given the size of the force and the distance moved.

- Explain how simple machines transfer energy in a way that offers an advantage.

- Recognise what is meant by temperature and how it is measured.

- Explain and make predictions about the direction of heat flow in different situations.

- Explain the difference between temperature and heat.

- Recognise that sound energy is transferred by waves and describe how sound waves are made in different situations.

- Explain how longitudinal waves carry sound. Relate the terms frequency and amplitude to sounds.

- Interpret and devise wave diagrams to represent different sounds of different wavelength and amplitude.

- Recognise an echo as a reflection of sound.

- Describe how to measure the speed of sound, and how the speed of sound can be used in different applications to measure distances.

- Use calculations to measure the speed of sound and the distance of objects in different applications, applying ideas about echoes.

- Recognise that some materials are good at reflecting sound and others can absorb it.

- Use the particle model to explain why sound cannot travel through a vacuum. Explain what is meant by reflection and absorption of sound.

- Use the particle model to explain why the speed of sound is different in solids, liquids and gases, and how energy is transferred in the reflection and absorption of sound.

- Recognise that different organisms hear differently. Name different parts of the human ear.

- Explain how parts of the ear are adapted to enable us to hear. Describe what is meant by the term hearing range.

- Compare and contrast the detection of sound by an ear and a microphone.

- Describe what is meant by infrasound and ultrasound.

- Describe a wide range of applications for ultrasound and infrasound.

- Explain why these waves are suitable for their applications.

Questions

Questions 1–7

See how well you have understood the ideas in the chapter.

1. Which of the following is the unit of energy? [1]

   a) kilogram   **b)** kilojoule   **c)** kilometre   **d)** kilohertz

2. Which of the following is not a fuel? [1]

   a) petrol   **b)** hydrogen   **c)** coal   **d)** air

3. What is the frequency range of ultrasound? [1]

   a) below 20 Hz **b)** between 20 and 20 000 Hz **c)** above 20 000 Hz **d)** there is no range

4. Describe the energy transfer when a ball falls from a height. [2]

5. Describe *two* uses of ultrasound. [2]

6. State *two* ways that energy can be stored. [2]

7. Describe how to measure the speed of sound. [4]

Questions 8–13

See how well you can apply the ideas in this chapter to new situations.

8. Julia's science teacher tells her that 'energy-efficient' light bulbs are better to use because they waste less energy as heat. But Julia knows that her mother, who is a farmer, uses old-fashioned filament light bulbs to keep newly hatched chicks warm in winter. Which of these statements is correct?

   a) Julia's teacher is right – bulbs that transfer most of the energy as heat are always wasteful.

   b) The chicks don't need heat – they just need to see where they are going.

   c) Heat is only wasteful if you don't make use of it.

   d) Julia's mother should switch to energy-efficient light bulbs.

9. Look at the different waves shown in 1.6.20a. Wave (a) represents a note played in the middle of a piano. Which wave best represents a siren? [1]

   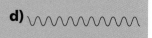

FIGURE: 1.6.20a

10. Emily's family are moving house. Their lounge is empty, with no curtains, carpets or furniture, and it sounds 'echoey'. Which of these statements is correct?

    a) Hard surfaces are good at absorbing sound.

    b) Sound travels faster than light.

    c) Sound travels faster in an empty room.

    d) Soft surfaces such as curtains are good at absorbing sound.

11. A bat sends out an ultrasound signal. It receives an echo just 0.5 seconds later. How far away is its prey? (distance = speed × time; the speed of sound in air is 330 m/s) [2]

12. Describe the energy transfers that occur as a waterfall drives an electricity generator. [2]

13. Figure 1.6.20b shows two different types of lever being used to lift the same load through the same distance.

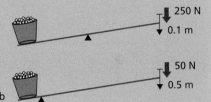

    a) Calculate the work done in each case.

    b) Which lever makes the work easier? Explain. [4]

FIGURE 1.6.20b

Questions 14–15

See how well you can understand and explain new ideas and evidence.

14. You are looking for the best possible fuel source for the future. Use the data in the table to make your choice. Give reasons for your answer. [2]

Fuel	Energy per gram (J/g)	State	Harmful products of combustion	Availability
coal	24	solid	carbon dioxide, soot, acid rain	running out
hydrogen	123	gas	none	plenty
petrol	46	liquid	carbon dioxide	running out
biofuel	33	liquid	carbon dioxide	renewable

15. The graph shows how people of different ages have reduced ability to hear certain frequencies. Use your knowledge of the ear and the data in Figure 1.6.20c to prove or disprove the statement 'Hearing loss gets worse as we get older'. [4]

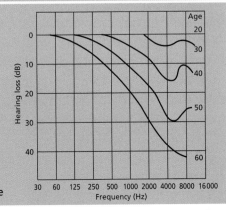

FIGURE 1.6.20c: Hearing loss with age

Glossary

absorb take in, for example energy from sound

accurate very close to the true value

acid substance that has a pH lower than 7

acid rain rainwater that is made acidic by pollutant gases

action force a force acting on an object that gives rise to an equal and opposite reaction force

adaptation gradual development that helps survival

adapted having special features that help survival

air pollution harmful particles and gases in the air

air resistance frictional resistance when something moves through the air

alkali substance with a pH higher than 7

alloy mixture of two or more metals

alveoli (singular alveolus) where gas exchange occurs in the lungs

amnion fluid that protects a foetus in the uterus

amoeba one type of unicellular (single-celled) organism

amplitude maximum displacement of a point on a wave from its undisturbed position

amylase enzyme that digests starch

analyse examine something to try to explain it

anaemia deficiency disease caused by lack of iron in the diet

anther pollen-producing part of the stamen of a flower

antibiotic drug that kills bacteria

anticlockwise turning in the opposite direction to clock hands

apparatus equipment for a scientific experiment

asthma disease affecting the breathing system

atom basic 'building block' of an element that cannot be chemically broken down

atomic mass number related to the mass of one atom of an element

atomic number number of an element in the Periodic Table

auditory nerve nerve that transmits signals from the ear to the brain

average speed speed calculated over the complete distance travelled

bacteria simple unicellular (single-celled) organisms, some of which can cause illness

balanced diet intake of foods that provide the correct nutrients in the correct proportions

balanced forces forces on an object that act in opposite directions and are equal in size

base (in chemistry) an oxide of a metal

body mass index (BMI) measure of whether a person is over- or underweight

boiling point temperature at which a liquid changes state to a gas

brine salty water

brittle easily cracked or broken by hitting or bending

Bunsen burner controllable gas flame used to heat substances

burning see *combustion*

capillary small blood vessel

carbohydrate food group including starches and sugars

carbon dioxide a gas in the air that plants use to make food, that animals breathe out, and that is produced by burning some fuels

carbonate type of compound containing carbon and oxygen

carpel female part of a flower

catalyst substance that speeds up a chemical reaction

cell 'building block' that all living things are made from

cell division when a cell divides in two

cell membrane layer around a cell that controls substances entering and leaving the cell

cell wall tough outer layer of plant cells

cervix opening between the uterus and the vagina

chemical digestion breakdown of food by enzymes in the digestive system

chemical energy see *chemical potential energy*

chemical formula chemical symbols and numbers that show which elements, and how many atoms of each, a compound is made up of

chemical potential energy energy stored in bonds between atoms, which becomes available when atoms are rearranged in a chemical reaction

chemical reaction a process in which one or more substances are changed into others, by their atoms being rearranged

chemical symbol abbreviation used to represent an element

chloroplast structure in plant cells where photosynthesis takes place

chromatogram pattern of results obtained in chromatography

chromatography process used to separate soluble substances

cilia tiny hair-like structures that help to keep mucus and dust out of the lungs

circulatory system the heart and blood vessels that transport essential substances around the body in blood

clockwise turning in the same direction as clock hands

cochlea part of the inner ear where vibrations are changed to electrical signals

collide hit against

combustion reaction of fuels with oxygen that produces heat

colony collapse disorder (CCD) widespread disorder in bee populations

compare look for similarities and differences

competition struggle between different organisms for survival

compound two or more elements that are chemically joined together, such as water (H_2O)

compress squash with a force

concentration amount of something per unit volume – for example sugar in water

conclusion opinion reached after studying all the evidence

condense turn from a gas into a liquid – as in water vapour condensing to liquid water

conductor (of heat) material that allows heat to pass easily

contact force force arising when objects are touching – for example friction

control variable factor kept constant in an investigation

counterweight a weight on a structure such as a crane, which balances a load to avoid toppling

crude oil black liquid extracted from the Earth from which petrol and many other products are made

crystal solid having a regular structure

crystallisation when a solvent evaporates to leave crystals of solid

current (electric) flow of charge in an electric circuit, which transfers energy

cytoplasm the main component of cells

data information, often in the form of numbers, obtained from surveys or experiments

decibel (dB) unit of sound loudness

deficiency disease illness caused by lack of a certain nutrient

deformation change of shape

degree Celsius (°C) unit on a temperature scale

density mass of a material per unit volume

dependent variable variable that is measured in an investigation

diaphragm muscular layer at the base of the chest cavity

diatomic description of a molecule containing two atoms

diet what a person eats over a period of time

diffusion particle movement that causes particles in a liquid or gas to spread out evenly

digestion breakdown of food in order to obtain energy

digestive system group of organs that together enable digestion of food

dissolve when a solid mixes with a liquid so that it can no longer be seen

distance length of the path covered on a journey

distillation process for separating liquids by evaporating them, then condensing the vapours

drag resistive force of air or water

ductile able to be stretched out a lot

dynamo device that transfers energy of movement as energy in electricity

ear canal where sound waves enter the ear

ear defenders protection worn for the ears when near loud sounds

ear drum membrane in the ear that transfers energy from sound waves to the inner ear

echo reflection of a sound wave

efficient not wasteful of energy

effort force applied when using a machine

egg cell female sex cell in animals

elastic behaviour when a material returns to its original shape and size after a deforming force is removed

elastic limit maximum force that can be applied for a material to remain elastic

elastic material see *elastic behaviour*

elastic potential energy energy stored when an elastic object is stretched or compressed

electrical energy energy of an electric current in a circuit

electrolysis chemical process that involves using electricity

element substance made up of only one type of atom

embryo young foetus before its main organs are formed

emit transfer energy away from an object

endometriosis condition in the reproductive system of a woman that may cause infertility

energy something has energy if it has the ability to make something happen when that energy is transferred

energy transfer diagram arrows that show how energy is transferred from one form to other forms

enzyme substance that enables a chemical process in the body

equation (chemical) a chemical reaction written in terms of reactants and products

equivalent the same as

euglena one type of unicellular (single-celled) organism

eukaryote one of a group of unicellular organisms that have a nucleus

evaluate study information and make a judgement about it

evaporate turn from a liquid to a gas – such as when water evaporates to form water vapour

evidence information gathered in a scientific way, which supports or contradicts a conclusion

extension amount by which an elastic material has got longer

extract ion get out – for example metal from rock

faeces solid body waste released through the anus

fat body tissue that acts as an energy store; a food group

fermentation a type of respiration in micro-organisms such as bacteria

fertilisation when a male sex cell fuses (joins with) a female sex cell

fibre (in food) component of food needed to avoid constipation

filament (of flower) 'stalk' of the stamen that supports the anther

filter material with microscopic holes used to remove insoluble solids from liquids

filtration separation of a solid from a liquid using a filter

flagellum 'tail' on a unicellular organism that helps it to 'swim'

flammable burns easily

flora (gut) bacteria in the intestines

foetus an unborn baby

food group type of food, such as protein, needed for certain body processes

force a push, pull or turning effect

force multiplier machine designed to allow a small effort to move a large load

forensic term used to describe scientists who analyse evidence to help the police investigate crimes

formula see *chemical formula*

fossil fuel coal, natural gas and crude oil that were formed from the compressed remains of plants and other organisms that died millions of years ago

fractionating column tall tower in which fractional distillation is carried out at an oil refinery

frequency number of waves passing a set point, or emitted by a source, in a second

friction a force that opposes movement

fruit ovary of a plant after fertilisation – it contains seeds

fuel material that is burned for the purpose of generating heat

fulcrum point about which something turns – also called a pivot

gas one of the states of matter

germination when a seed begins to grow into a plant

global warming increase in the Earth's temperature due to increasing amounts of carbon dioxide in the atmosphere

glucose a simple sugar molecule

gravitational potential energy stored energy that an object has because of its height

gravity force that pulls masses towards one another

greenhouse gas gas in the atmosphere, such as carbon dioxide, that prevents heat transfer away from the Earth

group (chemical) vertical column of elements in the Periodic Table

gut digestive system

haemoglobin chemical in red blood cells that carries oxygen

halogen group of non-metal elements

hazard something that can cause harm

hazard symbol standard symbol that warns of a particular type of hazard

hearing range range of sound frequencies that an animal can hear

heat form of energy – energy is transferred as a flow of heat from a hot object to a cold object

Hooke's Law law that says that a spring will extend regularly as the force on it is increased – the extension is proportional to the load

horizontal parallel to flat ground

hydrocarbon compound, such as many fuels, containing only carbon and hydrogen

hydroelectric dam structure that allows the generation of electricity from water flowing downwards

hydrogen the lightest element

hypothesis idea that explains a set of facts or observations, and is the basis for possible experiments

immiscible liquids that do not mix, but form separate layers

inanimate non-living

independent variable a variable in an experiment that affects the outcome

inert chemically unreactive

infertility difficulty in becoming pregnant

infrasound sound with a frequency lower than 20 Hz

insecticide chemical applied to crops to destroy insect pests

insoluble unable to dissolve

intestines parts of the digestive system where substances are absorbed from food

joule (J) unit of energy

kilojoule (kJ) 1000 joules

kinetic energy energy that moving objects have

laboratory special room designed for science experiments

Law of Conservation of Mass in a chemical reaction, the total mass of the reactants is the same as the total mass of the products

lever simple machine that uses the turning effect of a force about a pivot

Liebig condenser apparatus used for distillation

limescale white substance composed of calcium compounds that were dissolved in water

limewater solution used to test for the presence of carbon dioxide

line diagram simple informative drawing

load weight to be lifted/moved

longitudinal wave wave in which the vibrations are parallel to the direction in which energy is transferred

loudness measure of the energy in a sound wave – measured in decibels (dB)

lubricant substance used to reduce friction

lungs main organs of the breathing system

magnetism effect of magnet poles on magnetic materials

malleable able to be bent without breaking

malnutrition health condition caused by insufficient intake of nutrients

mass amount of matter in an object – measured in kilograms (kg)

material anything made of matter

matter 'stuff' that the world is made of – its three states are solid, liquid and gas

measure use standard units and instruments to determine the size of something

melting point temperature at which a solid changes state to a liquid

meniscus curved surface of a liquid in a container

menstruation monthly breakdown of the uterus lining, leading to bleeding from the vagina (a period)

metalloid element that shows the properties of both metals and non-metals

microphone device for changing sound into an electrical signal

microscope optical device used to see magnified images of tiny objects and structures

mineral (in diet) element such as iron or calcium needed in the diet

mitochondria structures in a cell that produce energy

mixture two or more elements or compounds mixed together, but not chemically joined

model diagram or three-dimensional object that makes an idea easier to understand

molecule two or more atoms held together by strong chemical bonds

moment size of the turning effect of a force around a fulcrum (pivot), equal to the force × the distance from the fulcrum – it can be clockwise or anticlockwise

muscle cell specialised body cells that make up muscle tissue

native elements elements that occur naturally

nectar sugary liquid produced by flowers

nerve cell specialised body cell that transmits messages around the body in the nervous system

newton (N) unit of force

newtonmeter device that uses the stretching of a spring to measure force

nicotine addictive substance in cigarettes

noble gas unreactive gas in Group 18 of the Periodic Table

nucleus (of cell) part that contains the genetic material – it controls the cell

nutrient substance in food that we need to eat to stay healthy – such as protein

obesity medical condition in which the amount of body fat is so high that it harms health

objectivity freedom from bias in forming an idea

observation looking at something to discover features or changes over time

ore rock from which a metal is extracted

organ collection of tissues that work together to perform a function

organ system organs that coordinate with one another in body processes

organelle structure in a cell with a specific function (job)

organic compound compound that contains carbon

organism a living thing

ossicles tiny bones in the ear that amplify the vibrations of sound

ovary organ in female animals that makes egg cells; and in plants that contains ovules

oviduct tube in a female animal that carries the egg cell from the ovary to the uterus

ovulation release of an egg from the ovary

ovule female sex cell of a plant

oxidation chemical process that increases the amount of oxygen in a compound

oxide a compound made by reacting an element with oxygen

oxygen a gas in the air that is produced by plants, used by animals in breathing, and used in combustion (burning)

pancreas organ of the digestive system that produces enzymes

paper chromatography simple method of separating different inks or dyes

paramecium type of unicellular (single-celled) organism

particle very small part of a material, such as an atom or a molecule

penis sex organ of a male animal

period (menstruation) see *menstruation*

period (chemical) a horizontal row of elements in the Periodic Table

Periodic Table a table of all the known elements in order of atomic number

pesticide chemical applied to a crop to destroy pests

pH measure of acidity/alkalinity

photosynthesis process carried out by green plants – sunlight, carbon dioxide and water are used to produce glucose and oxygen

physical digestion mechanical breakdown of food in the mouth

pitch how high or low the frequency of a sound is

pivot point about which something turns – also called a fulcrum

placenta structure in a pregnant woman that provides nutrients to the baby

pollen male sex cell of a flower

pollen tube structure that grows from a pollen grain in order to fertilise an ovule

pollination process of transferring pollen from the anther of a flower to the stigma of a flower

polymer large molecule made up of a very long chain of smaller molecules

population number of organisms living in a particular area or habitat

precise when repeated readings of the same measurement give similar values

precision how close together, or spread out, repeated measurements are

predict suggest, with reasons, what may occur

premature baby born before it has fully developed

pressure force on a certain area

probiotic containing helpful bacteria

prokaryote one of a group of unicellular organisms that have no nucleus

protein food group important for growth and tissue repair

protozoa unicellular organisms such as amoeba and paramecium

puberty changes that occur to boys and girls as they become adults

pure containing only one type of element or compound

purifying making pure, or nearer to pure

radioactive describes a material that emits nuclear radiation

ratio link between two values – for example if the first value is twice the second value, the ratio is 2:1

reactants starting substances in a chemical reaction

reaction (chemical) see *chemical reaction*

reaction force when object A pushes with an action force on object B, B exerts an equal and opposite reaction force on A

reduction chemical process that reduces the amount of oxygen in a compound

reflect re-direct a wave, such as sound or light, often back towards the source

reflection when a wave, such as sound or light, bounces off a surface

reliable results of an experiment are reliable if the same results would be obtained again – even if carried out by different people using different equipment

reproductive system organs in an organism involved in producing offspring

respiration process in living things in which oxygen is used to release the energy from food

rickets deficiency disease caused by lack of vitamin D

risk the likelihood of a hazard causing harm

rock salt impure salt that is mined

root hair cell specialised plant cell on roots

safety reduced likelihood of harm or damage

saliva liquid produced in the mouth that helps swallowing and digestion

salt type of chemical compound – our table salt is sodium chloride

sample size number of things or people studied in a survey

Sankey diagram energy transfer diagram that shows the proportion of energy transferred as different forms

saturated (solution) when no more solute will dissolve

scan (medical) image produced, for example by ultrasound or X-rays, to check the health of internal body parts

scrotal sac protective covering of the testes

scurvy deficiency disease caused by lack of vitamin C

seed dispersal spreading of seeds from a plant to a new area

semen fluid in which sperm are carried

semiconductor material used in electronic circuits

semi-metal see *metalloid*

semi-permeable membrane allows some but not all particles to pass through

solubility mass of solute that dissolves in a solvent at a particular temperature

soluble able to dissolve (usually in water)

solute solid that dissolves

solution mixture formed when a solid dissolves in a liquid

solution mining method of extracting salt from rocks

solvent liquid in which something dissolves

sonar technique used on ships to measure the depth below the ship, by detecting reflected ultrasound

sonorous make a sound like a bell when hit

soundproofing using materials that absorb sound

speed how fast something travels – calculated using the equation speed (metres per second) = distance ÷ time

speed of sound the speed at which a sound wave travels

sperm cell male reproductive cell of an animal

sperm duct tube through which sperm travel from the testes

stable unlikely to topple or break up

stamen male part of a flower

starch large molecule made by plants as a form of food storage

starvation severe long-term shortage of energy from food

stem cells unspecialised body cells that can develop into other, specialised cells

stigma pollen-receiving part of a flower

stomach part of the digestive system where most food breakdown occurs

streamlined has a shape that helps an object to slip easily through water or air with very little friction

stretch use a pulling force to make something longer

style (of flower) female part of a flower through which pollen travels to fertilise an ovule

sugar sweet-tasting compound of carbon, hydrogen and oxygen – such as glucose or sucrose

surface area area of the outside surface of something

symbol see chemical symbol

tar chemical in cigarettes that causes lung disease

temperature the hotness of an object – measured usually in degrees Celsius (°C)

terminal velocity final constant speed that a falling object reaches

testes parts of a male animal where sperm are made

thermal decomposition a chemical change caused by heating when one substance is changed into at least two new substances

tissue collection of body cells that work together to carry out a task

toxic poisonous

transition metals metal elements in the middle of the Periodic Table

turbulence irregular changes in direction of flow

turning force force that causes a turning effect about a pivot

ultrasound sound with a frequency higher than 20 000 Hz

umbilical cord tissue that attaches a foetus to its mother's placenta

unbalanced forces forces on an object that do not cancel one another out

unit standard amount used to measure a physical quantity – for example metre, kilogram and second

urea waste product removed from the body by the kidneys

urethra in a male, a tube in the penis through which sperm travel in semen

uterus part of a woman's body where a foetus grows – also called the womb

vacuole bubble of water and nutrients in a plant cell

vacuum a space where there are no particles of matter

vagina part of a female through which a baby is born

valid a measurement is valid if it measures what it is intended to be measuring

vapour liquid that has evaporated

variable factor that may affect the outcome of an experiment

vertebrate animal with a backbone

vertically at right angles to flat ground – directly towards or away from the centre of the Earth

vibrate to-and-fro movement

vital capacity volume of the lungs

vitamin important nutrient needed in very small quantities

volatile evaporates quickly

water resistance frictional resistance when something moves through water

wave a regular vibration that transfers energy

waveform graph of the displacement of a wave motion, at different distances along the wave

wavelength distance along a wave from one point to the next point where the wave motion begins to repeat itself – for example crest to crest

weight force of gravity acting on an object

weightless when there is no mass nearby to exert gravity

work done energy transferred by a force moving a load – equal to the force × the distance through which the load moves in the direction of the force

yeast unicellular (single-celled) fungus

Index

Index

Index

Index

Index

Acknowledgements

Acknowledgements

The publishers wish to thank the following for permission to reproduce photographs. Every effort has been made to trace copyright holders and to obtain their permission for the use of copyright materials. The publishers will gladly receive any information enabling them to rectify any error or omission at the first opportunity.

(t = top, c = centre, b = bottom, r = right, l = left)

Cover and title page image © 1stGallery/Shutterstock

p 8 (t) Cliparea/Custom media/Shutterstock, p 8 (c) Maks Narodenko/Shutterstock, p 8 (b) Paul Prescott/Shutterstock, p 9 (t) Astrid & Hanns-Frieder Michler/Science Photo Library, p 9 (c) AMI images/Science Photo Library, p 9 (b) Eye of Science/Science Photo Library, p 8-9 Whatafoto/Shutterstock, p 10 Natural History Museum/Science Photo Library, p 11 Dejan Dundjerski/Shutterstock, p 12 Astrid & Hanns-Frieder Michler/Science Photo Library, p 15 (t) Steve Gschmeissner/Science Photo Library, p 15 (b) Dr Jeremy Burgess/Science Photo Library, p 15 (b) Dr Jeremy Burgess/Science Photo Library, p 21(tl) Science Photo Library, p 21 (tc) Geoff Tompkinson/Science Photo Library, p 21 Pr Michel Brauner, ISM/Science Photo Library, p 21 (b) A. Dowsett, Public Health England/Science Photo Library, p 24 (t) Pavel Bredikhin/Shutterstock, p 24 (b) Wildlife GmbH/Alamy, p 25 Tim Gainey/Alamy, p 26 Biology Media/Science Photo Library, p 28 Matteo Sani/Shutterstock, p 29 (l) Tyler Olson/Shuttertock, p 29 (r) Goghy73/Shutterstock, p 30 (t) Serge Vero/Shutterstock, p 30 (c) Vlada Z/Shutterstock, p 30 (br) Spaxiax/Shutterstock, p 30 (bl) Dario Lo Presti/Shutterstock, p 31 (l) Anna_Huchak/Shutterstock, p 31 (c) Kiorio/Shutterstock, p 31 (r) D. Kucharski K. Kucharska/Shutterstock, p 32 (t) Lu Mikhaylova/Shutterstock, p 32 (c) ZCW/Shutterstock, p 32 (b) AmnachPhoto/Shutterstock, p 33 Charles E Mohr/Science Photo Library, p 35 (t) Eye of Science/Science Photo Library, p 35 (b) Dr. Robert Markus, Visuals Unlimited/Science Photo Library, p 37 Eye of Science/Science Photo Library, p 38 Hasloo Group Production Studio/Shutterstock, p 40 Juergen Berger/Science Photo Library, p 42 P. Saada/Eurelios/Science Photo Library, p 43 Mauro Fermariello/Science Photo Library, p 48 Igor Dutina/Shutterstock, p 49 Jeff Rotman/Alamy, **p 49 Springer Medizin/Science Photo Library,** p 48-49 Christopher Meade/Alamy, p 50 Ifong/Shutterstock, p 51 (t) Marco Mayer/Shutterstock, p 51 (b) B. and E. Dudzinscy/Shutterstock, p 52 (t) Stephen Coburn/Shutterstock, p 52 (b) Science Photo Library, p 54 (t) WaveBreakMedia/Shutterstock, p 54 (bl) Paul Michael Hughes/Shutterstock, p 54 (b) MichaelJung/Shutterstock, p 54 (br) Rtimages/Shutterstock, p 55 Volodymyr Krasyuk/Shutterstock, p 56 Mauro Fermariello/Science Photo Library, p 57 Christine Osborne/Alamy, p 58 (t) Biophoto Associates/Science Photo Library, p 58 (b) Jeff Rotman/Alamy, p 59 (tl) Smit/Shutterstock, p 59 (tc) Melodia plus photos/Shutterstock, p 59 (tr) Ffolas/Shutterstock, p 59 (bl) Nataliia Pyzhova/Shutterstock, p 59 (bc) Hong Vo/Shutterstock, p 59 (br) Nattika/Shutterstock, p 60 Auremar/Shutterstock, p 62 GosPhotoDesignShutterstock, p 66 (l) Ulrich Mueller/Shutterstock, p 66 (b) Stu Porter/Shutterstock, p 68 Samuel Borges Photography/Shutterstock, p 70 Professors P. Motta & F. Carprino/University "La Sapienza", Rome/Science Photo Library, p 75 Getty Images, p 78 Andrew Lambert Photography/Science Photo Library, p 81 Andrew VU, Visuals Unlimited/Science Photo Library, p 86 (t) Monkey Business Images/Shutterstock, p 86 (tc) Berna Namoglue/Shutterstock, p 86 (bc) Patricia Hofmeester/Shutterstok, p 86 (b) Frannyanne/Shutterstock, p 87 (t) Caroline Green, p 87 (c) Andrew Lambert Photography/Science Photo Library, p 87 (b) Charles D. Winters/Science Photo Library, p 86-87 Dr Gary Settles/Shutterstock, p 88 Jacek Chabraszewski/Shutterstock, p 90 (l) Caroline Green, p 90 (r) Stocksolutions/Shutterstock, p 92 Monticello/Shutterstock, p 94 Anton Balazh/Shutterstock, p 95 Suzanne Porter/Alamy, p 96 (t) Dario Lo Presti/Shutterstock, p 96 (c) Patricia Hofmeester/Shutterstock, p 96 (b) Dream79/Shutterstock, p 98 Charles D. Winters/Science Photo Library, p 100 (l) LVV/Shutterstock, p 100 (r) Javier Trueba/MSF/Science Photo Library, p 102 (t) Charles D. Winters/Science Photo Library, p 102 (b) Malcolm Chapman/Shutterstock, p 104 Fotocrisis/Shutterstock, p 106 (t) Andrew Lambert Photography/Science Photo Library, p 106 (b) Adam Burton/Alamy, p 108 Shestakoff/Shutterstock, p 109 Dudarev Mikhail/Shutterstock, p 110 Charles D. Winters/Science Photo Library, p 111 Sinclair Stammers/Science Photo Library, p 112 Dennis Steen/Shutterstock, p 113 Isak55/Shutterstock, p 114 (t) Darrin Jenkins/Alamy, p 114 (c) Martyn F. Chillmaid/Science Photo Library, p 114 (b) HSE, p 115 Syda Productions/Shutterstock, p 122 (t) Feng Yu/Shutterstock, **p 122 (tc) Shebeko/Shutterstock,** p 122 (bc) Caroline Green, p 122 (b) XAOC/Shutterstock, p 123 (t) Agsandrew/Shutterstock, p 123 (b) Caroline Green, p 122-123 Ed Samuel/Shutterstock, p 124 (t) Triff/Shutterstock, p 124 (b) David Nunuk/Science Photo Library, p 128 Michal Vitek/Shutterstock, p 129 Asaph Eliason/Shutterstock, p 130 (t) Adisa/Shutterstock, p 130 (b) Aaron Amat/Shutterstock, p 131 (t) PHB.z (Richard Semik)/Shutterstock, p 131 (b) Charles D. Winters/Science Photo Library, p 132 (t) Charles D. Winters/Science Photo Library, p 132 (b) Rat007/Shutterstock, p 133 (t) Mike Flippo/Shutterstock, p 133 (b) Ho Philip/Shutterstock, p 134 Photo Researchers/Mary Evans Picture Library, p 135 (t) J. Palys/Shutterstock, p 135 (b) Ria Novosti/Shutterstock, p 136 (t) Science Source/Science Photo Library, p 136 (b) Karel Gallas/Shutterstock, p 137 Konrad Mostert/Shutterstock, p 138 (t) Charles D. Winters/Science Photo Library, p 138 (b) Denis Tabler/Shutterstock, p 140 (t) Caroline Green, p 140 (b) Peshkova/Shutterstock, p 142 Bright/Shutterstock, p 143 (l) Science Photo Library, p 143 (c) Martyn F. Chillmaid/Science Photo Library, p 143 (r) Andrew Lambert Photography/Science Photo Library, p 144 Shebeko/Shutterstock, p 145 Science Photo Library, p 146 (t) Andrey N Bannov/Shutterstock, p 146 (bc) Caroline Green, p 146 (br) Andrew Lambert Photography/Science Photo Library, p 148 (t) Elen_studio/Shutterstock, p 148 (b) Caroline Green, p 149 Clearviewstock/Shutterstock, p 150 (t) Christoffer Hansen Vika/Shutterstock, p 150 (b) Voronin76/Shutterstock, p 151 (t) Rvlsoft/Shutterstock, p 151 (c) Roadk/Shutterstock, p 151 (bl) Bonninstudio/Shutterstock, p 151 (br) Leonid Andronov/Shutterstock, p 152 Max Blain/Shutterstock, p 153 Andrew Lambert Photography/Science Photo Library, p 154 (t) Jiang Hongyan/Shutterstock, p 154 (b) J. Schelkle/Shutterstock, p 156 Marbury/Shutterstock, p 157 Charles D. Winters/Science Photo Library, p 161 Martyn F. Chillmaid/Science Photo Library, p 162 (t) Matejmm/iStock, p 162 (tc) Awe Inspiring Images/Shutterstock, p 162 (bc) Marcel Jancovic/Shutterstock, p 162 (b) Science Source/Sciene Photo Library, p 163 (t) Dorling Kindersley/Getty, p 163 (tc) Tomasz Trojanowski/Shutterstock, p 163 (bc) Max Earey/Shutterstock, p 163 (br) Igor.Stevanovic/Shutterstock, p 162-163 Dr Gary Settles/Shutterstock, p 164 Vitalii Nesterchuk/Shutterstock, p 165 (t) Markus Mainka/Shutterstock, p 165 (b) Pressmaster/Shutterstock, p 166 (t) Georgios Kollidas/Shutterstock, p 166 (b) Andrew Lambert Photography/Science Photo Library, p 167 Dimitar Sotirov/Shutterstock, p 168 Ioannis Pantzi, p 170 NASA/JPL/Science Photo Library, p 171 Garry Wade/Getty, p 172 Civdis/Shutterstock, p 173 (l) Andrew Lambert Photography/Science Photo Library, p 173 (c) Andrew Lambert Photography/Science Photo Library, p 173 (r) Andrew Lambert Photography/Science Photo Library, p 174 (b) Blazej Lyjak/Shutterstock, p 174 (t) Hurst Photo/Shutterstock, p 175 Ozaiahin/Shutterstock, p 176 VaLiza/Shutterstock, p 177 Maksim Toome/Shutterstock, p 178 (l) Maxisport/Shutterstock, p 178 (c) Maxisport/Shutterstock, p 178 (r) Creative Travel Projects/Shutterstock, p 179 Clive Tully/Alamy, p 180 Max Earey/Shutterstock, p 181 GermanSkydiver/Shutterstock, p 182 (t) Rich Carey/Shutterstock, p 182 (b) Serg_Dbrova/Shutterstock, p 183 Dr Gary Settles/Shutterstock, p 184 (t) Hamiza Bakirci/Shutterstock, p 184 (c) JennyT/Shutterstock, p 184 (b) Fotostory/Shutterstock, p 186 JeP/Shutterstock, p 187 WhiteHaven/Shutterstock, p 188 Ross Kinnaird/Getty, p 189 Tomasz Trojanowski/Shutterstock, p 190 (t) Brian A Jackson/Shutterstock, p 190 (b) Bryon Palmer/Shutterstock, p 191 Aflo Co. Ltd./Alamy, p 195 (l) Minerva Studio/Shutterstock, p 195 (r) Rob Bayer/Shutterstock, p 196 Sculpies/Shutterstock, p 202 (t) Racheal Grazias/Shutterstock, p 202 (tc) Val Thoermer/Shutterstock, p 202 (bc) Sirikorn Techatraibhop/Shutterstock, p 202 (b) Steve Allen/Science Photo Library, p 203 (t) Radu Razvan/Shutterstock, p 203 (tc) Tyler Olson/Shutterstock, p 203 (bc) Francois Loubser/Shutterstock, p 203 (b) Four Oaks/Shutterstock, p 203-204 Andrew Lambert Photography/Science Photo Library, p 204 (t) Cordelia Molloy/Science Photo Library, p 204 (c) Neamov/Shutterstock, p 204 (b) Neamov/Shutterstock, p 206 (t) Paul Brennan/Shutterstock, p 206 (bl) Germanskydiver/Shutterstock, p 206 (br) CroMary/Shutterstock, p 208 (t) Tor Eigeland/Alamy, p 208 (c) Design Pics Inc./Alamy, p 208 (b) European Press Association/Alamy, p 210 (t) Kaponia Aliaksei/Shutterstock, p 210 (c) Awe Inspiring Images/Shutterstock, p 210 (b) Cordelia Molloy/Science Photo Library, p 211 WhiteTag/Shutterstock, p 212 (t) Ian Cumming/Science Photo Library, p 212 (c) 2happy/Shutterstock, p 212 (b) BunnyPhoto/Shutterstock, p 216 DBImages/Alamy, p 217 (t) Jan Martin Will/Shutterstock, p 217 (b) Tomasz Bidermann/Shutterstock, p 218 (t) Francois Loubser/Shutterstock, p 218 (b) Pius Utomi Ekpei/AFP/Getty Images, p 222 Sirikorn Techatraibhop/Shutterstock, p 224 EBFoto/Shutterstock, p 226 (t) Panda3800/Shutterstock, p 226 (b) Galyna Andrushko/Shutterstock, p 228 Ethan Daniels/Shutterstock, p 230 MichaelJung/Shutterstock, p 231 Raphael Daniaud/Shutterstock, p 234 EcoPrint/Shutterstock, p 236 GE Medical Systems/Science Photo Library, p 237 Monty Rakusen/Science Photo Library.